Canadian Association

BUILDERS' MANUAL

R-2000

The CHBA Builders' Manual is published by the Canadian Home Builders' Association. It represents the leading edge of housing technology in Canada.

Since the early 1980's, CHBA, in conjunction with the R-2000 Program, has actively promoted the design, development and construction of energy efficient housing. The Builders' Manual now reflects more recent inclusion of environmental issues into programs such as R-2000 and Advanced Houses.

This edition of the CHBA Builder's Manual has received extensive review from individuals across the country. Researchers, building scientists, regulators and builders have provided valuable input in the revisions to the Manual. We are grateful for their assistance and their continuing commitment to the advancement of Canadian housing technology.

The Steering Committee

Ross Monsour, CHBA, Project Manager
Sandy Kaczmarek, CHBA, Project Assistant
Keith Hornsby, CHBA
Paul Gravelle, CHBA
John Broniek, CHBA/R-2000
Sherif Barakat, National Research Council
Terry Marshall, Canada Mortgage and Housing Corporation
Roger Henry, Natural Resources Canada/CANMET
Gary Sharpe, Natural Resources Canada/R-2000
Dave Finn, IRC/NRC
Ron Armstrong, Ontario New Home Warranty Program

CHBA Sub-Committee on Technical Advice and Publications

Contributors

Dara Bowser, Bowser Technical Inc.
Alex McGowan, Enermodal Engineering Ltd.
Skip Hayden, Natural Resources Canada/CCRL
Chris Mattock, Habitat Design + Consulting Ltd.
Tex McLeod, McLeod and Associates
Jacques Rousseau, Canada Mortgage and Housing Corporation
Rowena Moyes

Cover Design

Working drawing contributed by Gary Schaefer

Production

REIC Ltd.

ISBN 0-86506-054-1

Canadian Home Builders' Association
150 Laurier Avenue West, Suite 200
Ottawa, Ontario
K1P 5J4

TABLE OF CONTENTS

PART 2 — BUILDING ENVELOPE

Chapter *1*

INTRODUCTION

Canadians are recognized for building some of the best, most innovative, and most resource efficient housing in the world. For more than fifty years the Canadian Home Builders' Association (CHBA), in conjunction with the Canadian government, has worked with manufacturers, suppliers and researchers to provide builders with information about state-of-the-art housing technologies, systems and processes. This manual represents the latest initiative of the Association to inform Canadian builders, as well as practitioners in other countries, of construction techniques resulting in housing which is better built, more comfortable, more energy efficient and more environmentally sensitive than ever before.

The CHBA Builders' Manual has drawn upon the first hand experiences of leading Canadian builders, researchers and manufacturers in identifying practical and proven home building techniques. Many of the housing innovations introduced over the last decade are the result of partnerships between CHBA and government, including the R-2000 Program, the Advanced Houses Program and the Healthy Housing Program. The federal government's technical and financial support of these leading edge initiatives has gone a long way toward maintaining the reputation of Canadian home builders as being amongst the best in the world.

Such a reputation does not come easily. Quality building requires care, attention to detail, and the application of the most efficient and up-to-date construction techniques and technologies. As every good builder knows, in Canada's demanding climate, this attention to quality is essential.

This manual summarizes the basic principles and techniques of leading edge home building in Canada. It is not intended to be an exhaustive treatment of building science or design, nor is it a building code. Rather, it outlines the fundamental principles of building high quality, energy and resource efficient homes with enough detail to show how they can be applied to the houses that you are currently building. Finally, the manual takes a look forward at some of the trends and technologies which are likely to influence new home construction through to the end of the decade.

We are confident you will find this manual useful in designing and building marketable, quality housing. If you take the time to read it thoroughly, it will help you keep ahead in today's competitive market.

1.1 THE EVOLUTION OF CANADIAN HOUSING TECHNOLOGY

Over the past fifty years, there have been many changes to the way houses are built in Canada. Changes and innovations have resulted in a better product, consisting of higher quality materials and systems. Changes have also resulted in increased productivity on the construction site, allowing a quality product to be built in a shorter period of time. Homes today are filled with equipment, appliances and systems which had not even been considered fifty years ago. The host of kitchen appliances common to today's housing, sophisticated controls and security systems, as well as more advanced, energy efficient mechanical systems are but a few of the innovations which provide a new level of comfort, convenience and energy efficiency to today's home purchaser. In addition, the integration of on-site innovations such as power tools, sheet materials, and prefinished materials has dramatically improved production efficiency in the construction process.

Figure 1.1, adapted and extended from Canada Mortgage and Housing Corporation's booklet *Housing in Canada, 1946 to 1986 — An Overview and Lessons Learned* identifies some of the technological changes that have occurred in the industry since World War II. In general, these changes have increased the speed and efficiency of home building in Canada by reducing the amount of labour required to construct a site-built house.

Process	Practice	
Basement	'40s	– concrete block and site-mixed concrete used with site-built board formwork, boards reused as wall and roof sheathing
	'60s	– transit-mixed concrete used with prefabricated formwork
	'80s	– little change, some use of preserved wood foundations
	'80s/90s	– insulated concrete forms
	'80s/90s	– free draining membranes and insulations
Wall Framing	'40s	– platform frame, some stationary assembly line process, little use of power tools or piecework trades
	'60s	– precut studs, stationary assembly line with sequencing of piece work produced by trades
	'80s	– new framing options developed but little change in process, some use of pre-fabricated panels, use of pneumatic nailing devices
	'80s/90s	– increased use of composite materials
Roof	'40s	– laid out and erected by skilled tradesmen
	'60s	– engineered, prefabricated trusses in general use
	'80s	– new truss designs available, little change in process
	'80s/90s	– increased use of clay or composite tiles
Wall and Roof Sheathing	'40s	– boards
	'60s	– fibreboard and plywood sheets
	'80s	– waferboard sheets, insulative sheathings, use of pneumatic nailing tools
	'80s/90s	– common acceptance of insulative sheathings
	'80s/90s	– common use of house wrap air barrier materials
Siding	'40s	– wood clapboard, brick, stucco
	'60s	– precoated aluminum, steel and hardboard
	'80s	– vinyl siding
Plumbing and Heating	'40s	– site-fitted and installed
	'60s	– prefabricated chimneys, some ductwork subassemblies
	'80s	– plastic plumbing, prefabricated chimneys and flues, energy efficient equipment, mechanical ventilation systems, electronic air cleaners and power humidifiers
	'80s/90s	– broader acceptance of mechanical ventilation systems
	'80s/90s	– integration of components of mechanical systems
	'80s/90s	– broader use of ground source heat pumps
Interiors	'40s	– wet-finished with plaster, cured, brush-painted
	'60s	– dry-finished with drywall, roller-painted
	'80s	– little change, some use of spray-paint equipment
Windows, Doors and Cabinetry	'40s	– fabricated on-site
	'60s	– prefabricated windows, cabinetry, and countertops
	'80s	– sealed thermal windows, insulated metal doors, prehung doors, prefabricated stairs, modular cabinets
	'80s/90s	– super windows comprising low-E, gas fill, insulated spacer, low conductivity frames

Figure 1.1 *Examples of the Evolution in Housebuilding Technology*

In addition to increasing labour efficiency, many of the post-war changes affected the air leakage characteristics and thermal performance of houses. For example, the use of sheet sheathing materials rather than boards for walls and roofs has considerably reduced air flow through buildings. Thermal insulation was not commonplace in the 1940s, but by the 1960s, builders were using kraft-faced batts of insulation in walls and ceilings. Typical insulation levels at that time were RSI 2.11 (R-12) in ceilings and RSI 1.41 (R-8) in exterior walls.

However, energy efficiency was not a major concern for home builders or their customers until the oil price shocks of the mid- and late-1970s. As energy prices increased, Canadian builders began to look for ways to improve the energy efficiency of their product. In Saskatoon, researchers and builders developed "super energy efficient" homes that could be heated for only a few hundred dollars a year. These homes were similar in design and appearance to existing homes: the only visible differences were their thicker walls and heat recovery ventilation equipment.

In 1980, the federal government decided to stimulate the rate of construction of more energy efficient homes by establishing the Super Energy Efficient Home (SEEH) Program. Under this program, Energy, Mines and Resources Canada (EMR) provided builders with training and incentives to construct what are now called "R-2000" homes. These homes had to meet a certain standard in terms of energy efficiency and quality of construction. In 1982, the Canadian Home Builders' Association (CHBA) assumed responsibility for the delivery of the R-2000 Program. The program's main achievement includes providing technical training to builders across the country, and serving as a catalyst to the research and development of technologies which enhance the energy efficiency of all homes built across the country.

The construction techniques pioneered by the R-2000 program are now recognized as the leading edge of building quality, energy efficient housing for the North American climate. The details included in this manual all comply with good design principles based on a firm understanding of building sciences. Virtually all of the suggested details will comply with local building codes. It must be noted however, that the applicability of details and systems to specific codes will rest in the domain of local building officials.

More recently, the R-2000 and the Advanced Houses Programs of Energy, Mines and Resources Canada and the Healthy Houses Program of Canada Mortgage and Housing (all being delivered in conjunction with CHBA), are demonstrating new possibilities for resource efficient, environmentally sensitive housing. The Advanced Houses Project has established rigorous energy performance criteria—twice as stringent as the R-2000 requirements—and has instituted guidelines for electrical and water use both inside and outside of the home. The R-2000 Program is also adding environmentally sensitive design and construction practices to its performance requirements. These changes, as adopted at the provincial level, will maintain the project at the forefront of Canadian homebuilding. This manual summarizes the most up-to-date information on these proven techniques.

1.2 FEATURES OF QUALITY HOUSING

Houses built using the construction techniques described in this manual share several characteristics. These houses incorporate:

- cost-effective insulation levels
- a sealed building envelope
- high efficiency heating and cooling systems
- a continuously operating mechanical ventilation system, and
- efficient use of electricity, water and resources.

These houses have a number of attractive features.

Greater Comfort

An airtight building envelope reduces air leakage, eliminating uncomfortable drafts, and reducing dust, insects and most outside noise. At the same time, enhanced control of ventilation air and indoor humidity levels allows improved occupant comfort.

Healthier Indoor Environment

A continuously operating mechanical ventilation system exhausts stale air and replaces it with fresh outdoor air. The fresh outdoor air is mixed with indoor air, heated or cooled to a comfortable temperature, and distributed throughout the house.

The selection and use of materials with reduced toxicity and designs which minimize the potential for spillage of combustion gases are another means of ensuring a healthier indoor environment.

Better Energy Efficiency

Higher insulation levels and lower air leakage rates keep heat loss to a minimum. Efficient mechanical systems optimize the energy required for space conditioning and water heating. In addition, efficient electrical lighting systems and efficient appliances are a cornerstone to an energy efficient home.

More Durable Construction

Proper construction details reduce the moisture problems that can cause basement leakage and the premature deterioration of structural components and interior finishes.

Improved Resource Efficiency

Fixtures which reduce the use of water and electricity and construction processes based on the 3 R's (reduce, reuse and recycle) of waste management — in conjunction with energy efficiency — will all lessen the burden of the house on the broader environment.

1.3   THE HOUSE AS A SYSTEM

For hundreds of years, designers and builders considered each part of a building — foundation, floors, walls, roofs, windows and doors, plumbing and electrical systems, and heating, cooling and ventilation systems — individually and independently. Today, builders recognize that the various parts of a house work together as a system to create a comfortable, durable and energy efficient building. The house system itself interacts with the surrounding environment and with the occupants.

With the overall system, there are a number of sub systems at work at any one time:

* the external environment such as temperature, wind, air quality, dust, noise, etc.
* the occupants, including pets
* the building envelope including the foundation and above grade walls, windows, roof, and floors
* interior fixed components such as partitions, floors, and finishes
* appliances, equipment, and furniture, and
* the mechanical/electrical system including the heating, ventilating, air conditioning, plumbing, and electrical components.

The individual characteristics of these sub systems and the way they interact affect the performance of the house as a whole in terms of:

* the protection it provides from the external environment
* its links to the external environment such as views out and design contribution to the neighbourhood
* the usability of spaces and components within the home
* the quality of the interior environment including the balance of natural and artificial light, noise levels, air quality, temperature, humidity, dust levels, etc.
* the durability/longevity of the house and its sub systems
* overall energy performance
* capital and operating costs, and
* the character and ambience of the house.

The concept of a house as a system is best illustrated by an example. Supplying fresh air was once considered to be simply a matter of opening windows — an effective but highly inefficient approach. When mechanical exhaust systems and hot-air furnaces were introduced, they were regarded as totally separate components that had nothing to do with windows. This perception led to problems in some houses as the action of exhaust fans caused combustion gases to be drawn into living spaces.

Now we know that the interaction of a number of sub systems in a house affects indoor air quality. The tightness of the envelope determines the degree to which air infiltrates or exfiltrates. The cooking, laundry and washing habits of the occupants affect the production of moisture and the need for air exchange. Environmental factors such as wind speed and direction can affect heating and ventilation needs. All these factors influence the design and operation of the mechanical sub systems. The best way to provide comfortable, safe conditions for the occupants is through controlled ventilation.

1.4 HOW TO USE THIS MANUAL

The theme of the house as a system underlies the organization of this manual. As shown in Figure 1.2, a house is like a puzzle where each piece is one of the components of the system. Each chapter of this manual corresponds to a different piece of the puzzle.

Part 1 (Chapters 2 through 5) covers the fundamentals of building science and design. You should ensure that you have a good understanding of these chapters before proceeding further.

Part 2 (Chapters 6 through 11) deals with the building envelope: the foundation, floors, walls, ceilings and roof, and windows and doors. The chapters in this part of the manual describe various ways to ensure that the envelope is airtight and has adequate insulation levels. They also discuss the characteristics of windows and doors as they relate to the specification, selection and proper installation of good products.

Part 3 (Chapters 12 through 18) covers mechanical systems. We have attempted to provide enough information to help you decide what type of equipment is best suited to your needs, and to allow you to discuss mechanical systems intelligently with your subcontractors. Manufacturers and distributors can provide more detailed material on specific products or particular installations. Chapter 18 addresses trends towards integrated mechanical systems which are in the developmental stages.

Finally, **Part 4** (Chapters 19 and 20) addresses new trends affecting house design and construction through the 1990s. Chapter 19 focusses on environmental sensitivity through the discussion of electrical efficiency in appliances and lighting, water conservation, and waste management as they apply to the design and construction of the home. While not comprehensive, this chapter introduces some new concepts and provides direction on choices affecting the builder. Chapter 20 addresses developing technologies and housing design. This chapter forecasts trends that will affect housing design and construction: meeting the needs of an aging and physically challenged population; enhancing the affordability of new housing; integrating home automation systems; and moving more of the house construction process into the factory.

The manual also contains an index and several appendices. These include tables giving the permeability and insulating values of building materials and common conversion factors. There are also separate appendices for electricians, plumbers and drywallers. Finally, there is a short glossary of terms.

This manual is designed as a reference tool: it has been organized to deal with issues in one or two pages, where possible. We recommend that you read the entire book. We also urge you to keep it handy to help you solve any problems you might experience during construction. Finally, for more information on how to build better houses using the techniques described in this manual, contact your provincial Home Builders' Association. It can provide you with training and technical support to help you adapt the latest housing technology to your market's needs.

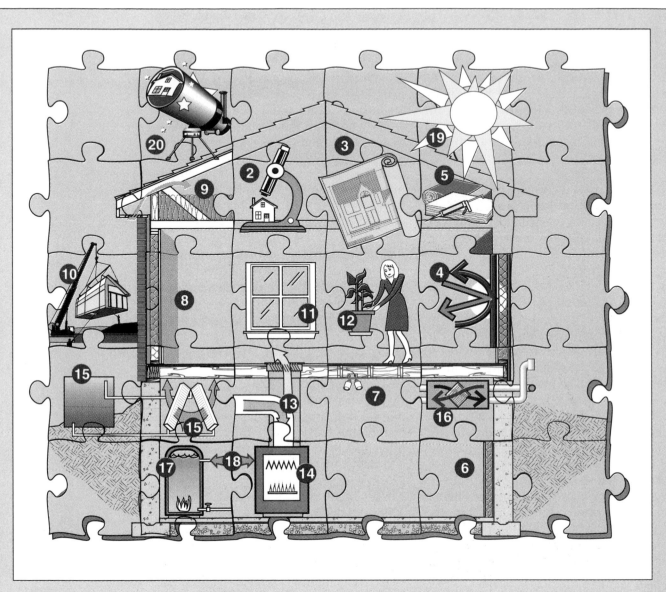

Figure 1.2   The House as a System

Part 1
BUILDING DESIGN

Chapter *2*

ASPECTS OF BUILDING SCIENCE

This chapter summarizes those key aspects of building science which affect the durability, comfort and energy efficiency of houses. These principles include heat, air and moisture flows. The chapter also introduces some basic concepts which should lead to an improved understanding of sound transmission and indoor air quality principles.

Take time to review this chapter carefully. These principles are the foundation of the construction techniques discussed in later chapters, understanding them is the first step towards eliminating common construction problems and building better houses.

2.1 CONTROLLING HEAT FLOW

2.1.1 Energy Flows in Houses

Our houses are balanced; the energy that enters a house is equal to the energy that leaves, usually in a different form than when it entered.

The following are ways energy can enter a building.

- **Purchased energy** is in the form of electricity, oil, natural gas, propane, or wood. This energy is used to heat and cool the house, to provide domestic hot water, and to operate lights and appliances.

- **Free solar energy** enters the house through glazed portions of the windows and doors. The contribution of solar energy to heating needs can vary considerably due to local weather conditions, the heating requirements of the building, and the size, orientation, shading and type of glazing units. Typically, 15 percent of an average home's heating is provided from free solar heat.

- **Internal heat gains** are derived from the occupants, lights and appliances, and can be an important energy source. These "free" gains reduce the need for space heating, especially during the early fall and late spring. In climatic zones that require space cooling, this contribution can become a factor in higher cooling costs.

The end uses of energy follow.

- **Space conditioning** is the heating or cooling required to maintain comfortable conditions within a building throughout the year. It usually represents the largest energy consumption in a single family dwelling. It is also the energy use most easily reduced by proper design, improved insulation and airtightness measures, and through the use of efficient mechanical equipment. In a well built home, space heating should consume only about one-third of the building's total purchased energy.

- **Water heating** will usually consume another one-third of the purchased energy. Savings on energy used for water heating can be obtained by controlling water temperature, increasing insulation levels on the domestic hot water tank, reducing hot water use, and installing higher efficiency equipment.

- **Lighting** generally accounts for a small amount of purchased energy — as little as 2 percent, or 1,000 kWh/yr, of the typical home's purchased energy requirements. However as heating, cooling and water heating requirements are reduced, lighting can assume a greater significance — to as much as 10 percent. This demand can be minimized by making use of natural light wherever possible and by installing more efficient fixtures in high use areas.

- **Appliances** include items such as stoves, refrigerators, washers, dryers, dishwashers and freezers. Typical use will result in consumption of 7,000 kWh/yr. The energy consumption of an appliance will depend on the model selected and the frequency of use. Some appliances entering the market consume 50 percent less energy than typical units.

2.1.2 Mechanisms of Heat Transfer

Heat naturally flows from a warm area to a relatively cooler area, for example, from the house interior to the exterior in winter, or from a warm room to a cool room or basement within the house. Heat flow can never be stopped, but it can be slowed down through the use of materials that have a higher resistance to heat flow, i.e., a higher R-value.

Heat is transferred by three different mechanisms (see Figure 2.1) which happen simultaneously.

• Heat transfer by **conduction** occurs when the molecules within a solid are excited by a heat source applied to one part of the object. The molecules in the object move about more vigorously, passing energy along the object.

• Heat transfer by **convection** occurs in *fluids* such as water or glycol or in gaseous materials such as air. As air or water is warmed, it becomes less dense and rises, causing cooler, denser air to be drawn in to replace it. Placing more dense insulation in empty stud cavities traps "dead" air which minimizes convection and slows heat transfer.

• With heat transfer by **radiation**, heat transfers from a warm object to a relatively cooler object by giving off heat waves. Even though there is normally air between two objects in a house, it will not be heated by these waves as they carry heat from one object to the other. For example, heat energy from the sun or a wood stove is in the form of radiant energy.

All three heat flow mechanisms are at work in a house at any given time. However, most heat is lost *through the building envelope, primarily by conduction and through the loss of heated air.*

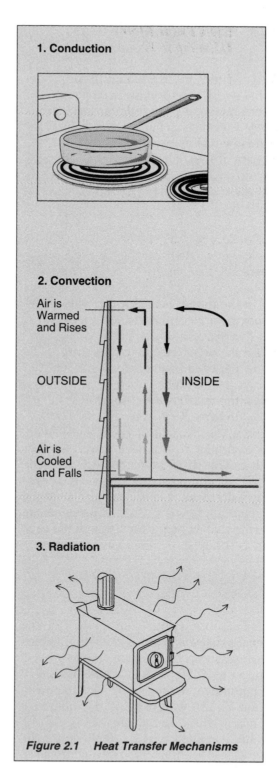

1. Conduction

2. Convection

Air is Warmed and Rises

OUTSIDE   INSIDE

Air is Cooled and Falls

3. Radiation

Figure 2.1   Heat Transfer Mechanisms

2.1.3 Basic Equation for Heat Transfer by Conduction

Heat transfer by conduction through the elements of a building envelope can be determined using the following equation:

heat flow (Q), is equal to
<u>the area of a surface (A) times the temperature differential (ΔT)</u> divided by the thermal resistance (RSI)

or

$$Q = \frac{A \times \Delta T}{RSI}$$

where:

Q   = the rate of heat flow, in watts (W);
A   = the area of the surface under consideration, in m²;
ΔT   = the difference in temperature between inside and outside, in °C; and
RSI = the resistance to heat flow, in m² x °C/W.

For example, if you have a wall section with an area of 10 m² and an RSI value of 2.1 (R12), and you know that the temperature indoors is +20°C and the outdoor temperature is -10°C, then you can calculate the rate of heat flow through the section as follows:

$$Q = \frac{A \times \Delta T}{RSI} = \frac{10 \times 30}{2.1} = 142.9 \text{ W}$$

This heat loss is equivalent to the amount of energy used to power a 150 W light bulb. A similar area (10 m²) of double glazed window at RSI 0.35 (R2) would lose 857.1 W, equivalent to the energy used to power almost six 150 W light bulbs (not including solar heat gain). Ten square meters (10 m²) of attic insulated to RSI 7.0 (R40) would lose 42.9 W, equivalent to the energy used to power a 40 W light bulb.

A comprehensive table of thermal resistance values for a wide range of materials is provided in Appendix I. This appendix also outlines how to calculate the overall thermal resistance of a wall system, which can also be applied to other envelope assemblies.

2.1.4 Air Leakage

Several methods can be used to estimate the air leakage rate of a building.

One method is to calculate the total combined length of all cracks in a building — around windows and doors, at the junction of the foundation and the header (ring) joist, etc. Using the ASHRAE Crack Method, it is possible to estimate the percentage of heat lost through air leakage. Needless to say this technique would represent little more than an educated guess.

It is also possible to express the overall air tightness level of a building as an "equivalent leakage area" (ELA). The ELA is defined as a single hole with a size or area equivalent to the total area of all of the cracks and holes in the building envelope. An ELA is theoretical.

A "blower door" can also be used to measure the air tightness of a building's envelope. This device consists of a fan, a calibrated opening used to measure air flow, an adjustable door, and metering equipment (see Figure 2.2). The blower door is installed in a door opening of the building that is to be tested and the fan is turned on, with the windows and doors closed, to blow air out of the building and depressurize the home. Operation of the fan will cause an equal amount of air to leak in through cracks and holes in the house envelope.

The amount of air leaving the building, typically at 10 Pascals (Pa) pressure, can be measured using the blower door apparatus. The measured air flow rate at 10 Pa, expressed in m³/hr, when divided by the building volume, expressed in m³, provides the airtightness value of the building, expressed in air changes per hour (ACH) at 10 Pa. To qualify as R-2000, for example, a house must have a maximum air leakage rate of 1.5 ACH @ 50 Pa.

The recent changes to the National Building Code requiring a "continuous" air barrier reflect a recognition that homes should be constructed in a manner to reduce air leakage for greater durability and comfort.

2.1.5   Summary

By using the R-values of the various building components, the air leakage rate, the efficiency of mechanical equipment, the averaged local climatic information, and estimates of internal and solar heat gains, it is possible to estimate the annual energy requirements of a given house.

Identifying the rate of heat loss allows sizing of the heating/cooling equipment. These calculations can be done manually, or with the aid of one of the many computer programs currently on the market, such as HOT2000 ® (available from CHBA). Further details are provided in the following chapter.

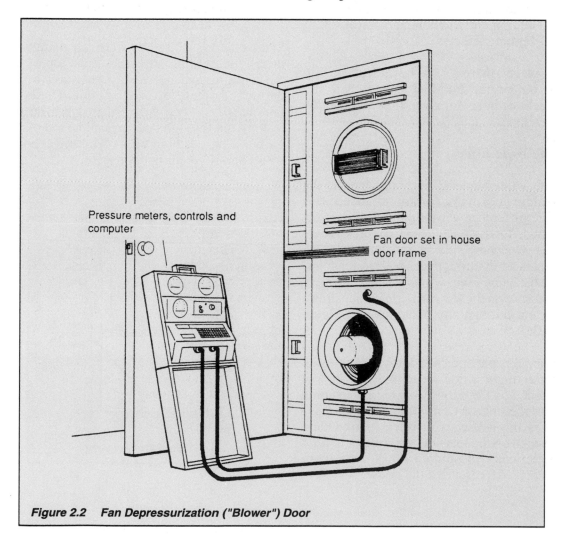

Pressure meters, controls and computer

Fan door set in house door frame

Figure 2.2   Fan Depressurization ("Blower") Door

2.2   CONTROLLING AIR FLOW

2.2.1   Factors Causing Air Flow

Much as heat always flows from hot to cold, air flows from a high (positive) pressure area to a low (negative) pressure area. The amount of air flow depends on the size of the openings and the pressure difference across the envelope.

Several factors cause a pressure difference from inside to outside, across the building envelope:

- wind
- temperature differences that cause convection i.e. the stack effect, and
- the operation of chimneys, ventilation equipment, and heating/cooling distribution systems.

However, there will be no air leakage into or out of the building unless there are cracks or holes for air to flow through the envelope.

The Wind Effect

The wind blowing over and around the building is a major contributor to the pressure difference across the building envelope. Winds create a positive pressure on the windward side of the house, which forces air in through any cracks and holes. At the same time, winds create a negative pressure on the leeward side, which draws air out through any openings (see Figure 2.3).

Many people open windows to get outside air at night when sleeping. A window opened on the windward side of the home provides lots of ventilation air and pressurizes the house. Conversely, a window opened on the leeward side results in little direct entry of outdoor air into the room and can depressurize the house.

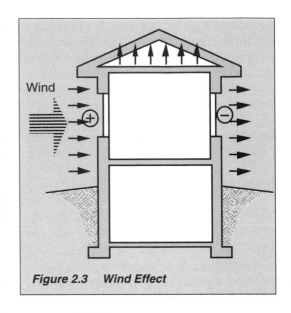

Figure 2.3   Wind Effect

The Stack Effect

Stack effect is caused by warm air rising, thereby increasing pressures at the top of the building and decreasing pressures at the lower extremeties of the building. These pressure differences will result in warm air leaking out through openings at the top of the building (see Figure 2.4). This effect is what makes hot air balloons rise.

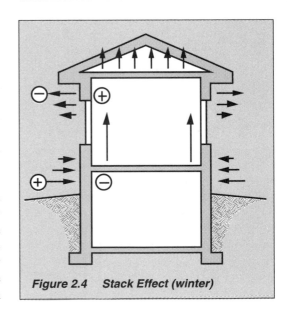

Figure 2.4   Stack Effect (winter)

Air tightening measures will reduce the size of the leakage opening, reducing air leakage rates, although the effect can never be completely eliminated. More floors in a home (for example, a three storey home versus a bungalow), increase the stack effect. Similarly, the greater the temperature difference from inside to outside, the greater the stack effect.

The Flue and Mechanical Effects

Chimneys and flues serving furnaces, boilers, water heaters, wood stoves and fireplaces also cause a pressure difference that results in air leakage. When heating equipment is operating, air is actively drawn out of the house through the chimney, venting the products of combustion. Even when the equipment is not operating, improperly functioning dampers can permit warm air to escape up the chimney. Chimneys, whether active or passive, create a negative pressure within the house which in turn causes outdoor air to be drawn through holes in the envelope (see Figure 2.5). In effect, active chimneys have, over the years, provided for ventilation in the home.

The mechanical effect, particularly through the operation of an exhaust appliance, is similar to the flue effect. Many houses contain mechanical equipment such as range hoods, bathroom fans, clothes dryers, central vacuums, and power vented hot water heaters which exhaust air to the outside. When these devices are operating they blow air out of the house, creating a negative pressure within the house. This change of pressure results in outdoor air leaking in through any cracks or holes in the lower levels of the envelope. Mechanical equipment may blow more air out of the home than the envelope can "leak" in: the resulting negative pressure can lead to "backdrafting" or "spillage" of combustion products from fuel-fired appliances.

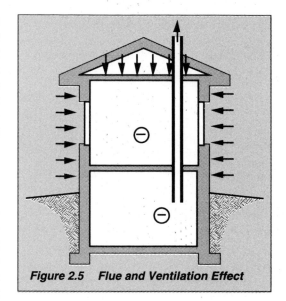

Figure 2.5   Flue and Ventilation Effect

For this reason, all combustion appliances must always have their own air supply and provision for all required make-up air. Sealed combustion equipment, i.e. direct vent, does not affect house pressures and when properly installed cannot leak combustion products into the home.

Recent research has yielded some information on the impact of ductwork that passes in and out of the house envelope or air barrier, into unheated spaces.

The leakage of conditioned air to the exterior through unsealed ductwork can create a significant negative pressure inside the building. The resulting infiltration of exterior air can affect energy performance, comfort, and the structural integrity of the envelope. Ductwork routed in and out of the air barrier should be avoided. If it is required, ensure that the ductwork is properly sealed at the joints and at the points where the ductwork penetrates the air barrier. In the United States, where this is a much more common practice, serious problems are being uncovered.

Figure 2.6 illustrates the effect of these forces acting simultaneously on a house, and the impact on the location of the neutral pressure plane discussed in the next section.

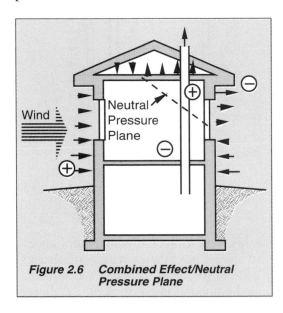

Figure 2.6   Combined Effect/Neutral Pressure Plane

The Neutral Pressure Plane (NPP)

The neutral pressure plane is a concept that can help to explain how air flows in and out of houses. The plane is an imaginary line where the pressure difference between inside and outside is zero. Below the NPP, air enters the building, while above the NPP, air leaks out of the building. The location of the neutral pressure plane will vary constantly depending on the wind effect, the stack effect and the effects of flues and mechanical equipment (see Figures 2.6 and 2.7).

If there are holes and cracks in the building envelope below the neutral pressure plane, cold outdoor air will be drawn into the living space (see Figure 2.7). This process is often the cause of uncomfortable drafts and dust infiltration. If the holes or cracks are above the neutral pressure plane, warm moisture-laden indoor air will leak out or exfiltrate. This process can lead to deterioration of the structure, problems such as peeling paint, efflorescence, deterioration of mortar, and spalling of brick work. In theory, air is neither leaking in nor out of the building at the neutral pressure plane.

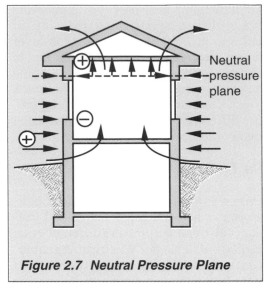

Figure 2.7 Neutral Pressure Plane

The concept of the neutral pressure plane can help to explain why houses with the same equivalent leakage area (see Section 2.1.4) may perform differently. Since the neutral pressure plane changes as constantly as its variables — winds, the operation of exhaust appliances and the stack effect — the best way to ensure good overall performance is to minimize the number of air leakage points throughout the building envelope by means of a continuous air barrier.

2.2.2   The Role of Air Flow in Heat Loss and Moisture Transfer

As the above discussion makes clear, air flow plays a critical role in both heat and moisture flows out of the building. The stack effect will cause warm moisture-laden air to exit from the higher levels of the house. As the moist air cools it can condense in walls and attics which can lead to premature deterioration of structural components. Air leakage can also account for 25 to 40 percent of the total heat loss of a house.

The way to avoid heat loss and moisture transfer due to air flow is to eliminate **uncontrolled** air exchange by sealing potential air leakage points during construction, and to substitute the required amount of **controlled** air exchange using mechanical ventilation.

2.3 CONTROLLING MOISTURE FLOW

2.3.1 The Three Phases of Water

Water has three phases:

- below 0°C (32°F) it exists as a solid
- between 0°C (32°F) and 100°C (212°F) it is a liquid, and
- above 100°C (212°F) it is a gas.

All three phases of water may be found in a building envelope during colder weather; however, there is water vapour in the air at all temperatures. Moisture as water vapour is not a problem, but water vapour, when cooled, condenses and can result in problems which cause conflict between builders and homeowners.

2.3.2 Mechanisms of Moisture Movement

Water can move from one place to another in four ways:

- by gravity
- through capillary action
- via air flow, and
- by diffusion.

Gravity

Gravity causes bulk or liquid water to seek its lowest point, travelling the path of least resistance. For example, if improper drainage directs surface run-off towards a basement wall, it may enter the building through a crack in the wall and cause a basement leak. Similarly, roof leaks, wet walls, etc., can be minimized by:

- the proper design and construction of roofs, walls and foundations to shed water
- the appropriate use of flashing, eaves-troughing, dampproofing and footing drains to control flows, and
- proper grading to direct water away from the house.

Capillary Action

Capillary action causes liquid water to move upwards, downwards or sideways through thin tubes in materials. A much more common example is a paper towel with one corner in a spilled liquid.

Capillary action affects foundations made of concrete or other porous materials. To the water in the ground, the pores in the concrete or mortar are like long, fine tubes. Water will rise via capillary action up the foundation wall until it is released into an area with a low water vapour pressure, usually inside the basement or crawlspace. This is often why the bottom few inches of a basement or crawlspace wall appear damp. Building a basement in sand is great for eliminating bulk water but provides no protection against the capillary rise of water.

If the water is prevented from diffusing into the basement, it will continue to rise until it can escape. If it is forced to rise to the top of the foundation wall, it can result in rot to the sill plate, headers and floor joists, particularly if these are embedded in the foundation.

There are two ways to minimize capillary action.

- Seal the pores in materials to prevent the entry of water by means of coatings or membranes.

- Make the pores large enough to discourage water entry — use a layer of coarse material, not sand or pit run, under concrete floor slabs.

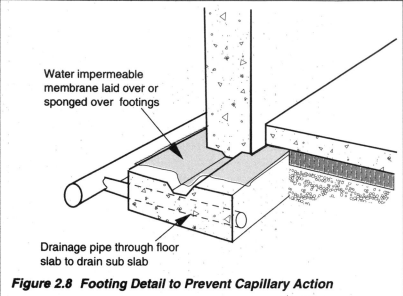

Water impermeable
membrane laid over or
sponged over footings

Drainage pipe through floor
slab to drain sub slab

Figure 2.8 Footing Detail to Prevent Capillary Action

Dampproofing foundation walls does not prevent water from rising through the footing. A water impermeable membrane or foundation coating should be installed over footings, before the foundation walls are poured, to prevent capillary action up the walls. The diagram also shows a through-the-footing drain tile that will help to reduce water pressure under the slab, reducing the potential for flooding in the basement or crawlspace (see Figure 2.8).

Air Flow

Moisture enters attics and wall cavities through air flow. Figure 2.9 illustrates potential air leakage sites where warm, indoor air can carry water vapour into the shell of the building. All of these sites must be sealed to prevent warm, moist air exfiltrating into the building envelope

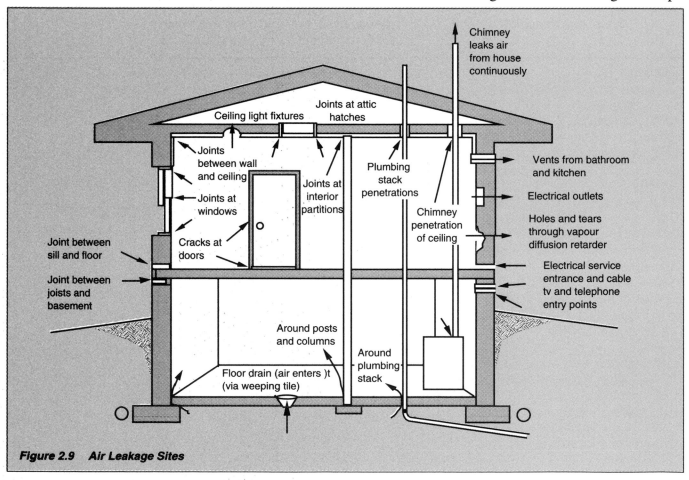

Chimney
leaks air
from house
continuously

Ceiling light fixtures

Joints at attic
hatches

Joints
between wall
and ceiling

Plumbing
stack
penetrations

Vents from bathroom
and kitchen

Joints at
interior
partitions

Joints at
windows

Electrical outlets

Chimney
penetration
of ceiling

Holes and tears
through vapour
diffusion retarder

Cracks at
doors

Joint between
sill and floor

Electrical service
entrance and cable
tv and telephone
entry points

Joint between
joists and
basement

Around posts
and columns

Around
plumbing
stack

Floor drain (air enters)t
(via weeping tile)

Figure 2.9   Air Leakage Sites

where condensation can occur. Recent changes to the National Building Code of Canada require that air barriers in the house be continuous. The airtightness standards of the R-2000 Program meet this requirement.

Diffusion

Diffusion is a process whereby water vapour passes through a seemingly solid material. Just as heat flows from a warm area to a cool area, water vapour will seek to flow from an area of high vapour pressure to an area of low vapour pressure.

Since some materials slow the diffusion of moisture better than others, the Building Code has established certain performance requirements for vapour retarders. A list of the water vapour transmission coefficients of various materials is included in Appendix II. Since vapour diffusion is never stopped, merely retarded, this publication uses the term vapour diffusion retarder (VDR) in place of vapour barriers. To ensure water vapour does not reach a temperature where it will condense, the VDR must be placed on the warm side of the wall.

If there was no air pressure difference causing air flows, there would be no concern for air leakage, but recent studies have shown that approximately 100 times more moisture enters the building structure by air flow than by diffusion (see Figure 2.10). Good building practice and codes require that insulated ceiling, wall and foundation assemblies incorporate a VDR in the assembly.

2.3.3   Sources of Moisture in Houses

Since warm air can hold more water vapour than cold air, moisture levels are considerably higher inside a heated house than outside under normal winter conditions.

- Moisture is generated by day-to-day activities such as breathing, cooking, bathing, washing and drying clothes.

- Furniture, drywall, and the framing materials inside the house absorb moisture from humid air during the summer. This is called "seasonal storage". Once the heating season begins, this moisture is released into the house air (see Figure 2.11).

- Moisture also enters the house from the concrete in basement and crawlspace foundations and from the floor slab. If the crawlspace or basement of an older house has an earthen floor, this can be the biggest single source of moisture.

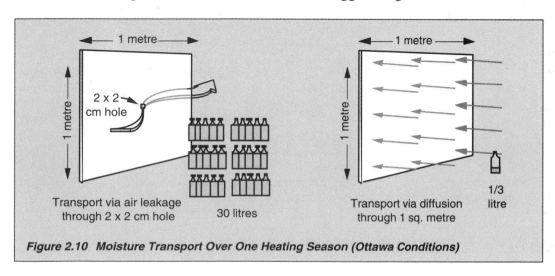

Figure 2.10 Moisture Transport Over One Heating Season (Ottawa Conditions)

• Firewood stored in the house to dry can also give off considerable moisture.

• New construction has considerable moisture stored in the materials used to construct the home. Concrete, drywall, and framing lumber, for example, contribute considerable moisture in the first year or two.

Figure 2.12 shows the amounts of moisture that various activities contribute. In a well built home, indoor humidity levels are controlled by a combination of, better building practice and the use of a properly designed and operated mechanical ventilation system.

The psychrometric chart (see Figure 2.13) can be used to determine when condensation will occur on windows or on other cold surfaces.

The thermal resistances of the wall, window, or ceiling assemblies, and the inside and outside temperatures, will provide the information necessary to plot the thermal gradient across the assembly (see Figure 2.14).

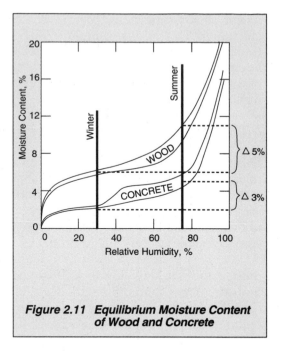

Figure 2.11 Equilibrium Moisture Content of Wood and Concrete

Using the psychrometric chart and the thermal gradient, it is possible to determine the location of the "dew point", the point at which condensation occurs, within a building assembly. If you do this calculation, you will be able to ensure that the VDR is always on the warm side of the dew point.

TYPICAL MOISTURE SOURCES PER DAY			
Occupant Related		**Building Related**	
• 4 occupants	5 litres (1.3 gal)	• seasonal building storage i.e. framing, drywall, concrete	8 litres (2.1 gal)
• clothes drying indoors	1.2 litres (0.32 gal)	• exposed uncovered earth crawlspace	40-50 litres (10.5 - 13.2 gal)
• floor washing/10m³ (107 sq. ft.)	1 litre (0.26 gal)	• drying and burning of firewood	5 litres (1.3 gal)
• cooking (3 meals/day)	1 litre (0.26 gal)	• new construction – drying of framing and concrete over first 18 months	4-5 litres (1 - 1.3 gal)
• dishwashing (3 meals/day)	1/2 litre (0.13 gal)		

Figure 2.12 Typical Sources of Moisture in Houses

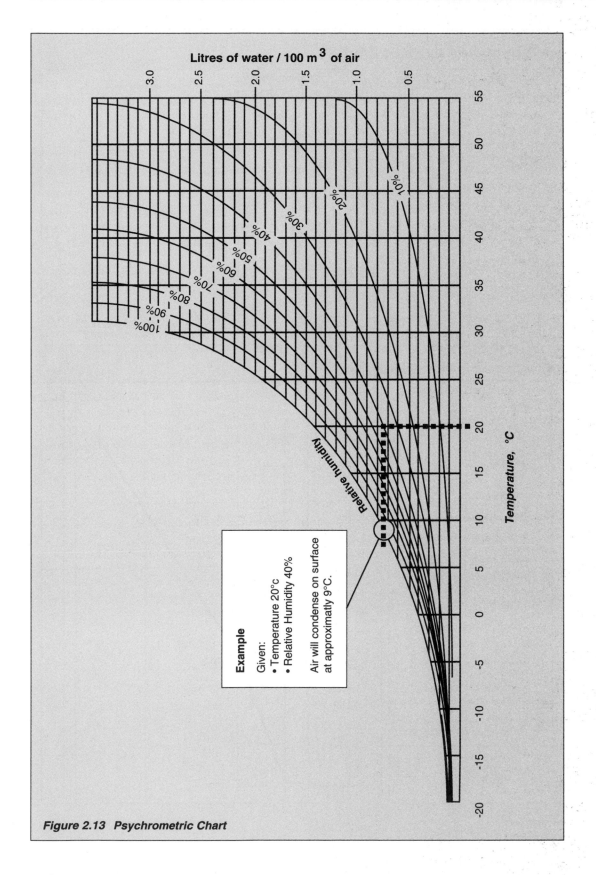

Litres of water / 100 m³ of air

Figure 2.13 Psychrometric Chart

Example

Given:
- Temperature 20°c
- Relative Humidity 40%

Air will condense on surface at approximatly 9°C.

Temperature, °C

Relative humidity

Actual Example of Temperature Conditions Through Test Wall Cavity

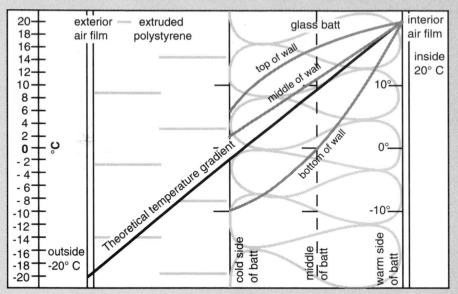

TEST DESCRIPTION
- No air barrier or VDR
- Ambient conditions:
  cold side: - 20°C
  warm side: + 20°C
  total $\triangle T_t$ 40° C
- Test duration: 300 hours

WALL COMPOSITION
- extruded polystyrene RSI 1.74
- glass batt   RSI 2.11
- films   RSI 0.15
- total   RSI 4.00

Actual measured
temperature ————

Calculation of Theoretical Temperature Gradient Through A Wall

	Wall Component	RSI	$\triangle$T across component (RSI/RSI t x $\triangle$Tt)
1	exterior air film	0.0053	0.039
2	metal siding	0.0211	0.157
3	air space with strapping	0.0300	0.224
4	50mm (2") extruded polystyrene	1.761 (R10)	13.125
5	38 x 89mm (2x4") studs with batt insulation	2.113 (R12)	15.749
6	38 x 64mm (2x3") horizontal strapping with batt insulation	1.409 (R8)	10.502
7	gypsum wall board	0.0062	0.046
8	interior air film	0.0211	0.1572
	Totals	RSI$_t$ 5.3667	$\triangle T_t$ 40° C

Winter conditions

Inside temperature	+20° C
Outside temperature	-20° C
$\triangle$ T$_t$ across wall assembly	40° C

Figure 2.14 Temperature Gradient

2.3.4   Damage Potential

Water vapour by itself is not a problem, but it can cause problems when it reaches the dew point temperature and condenses. When water condenses on surfaces, it can cause aesthetic problems and break down protective coatings of paint or caulking. If it condenses within the envelope, it will reduce the performance of the insulation. If allowed to accumulate within the envelope, it can lead to more serious problems: if the moisture content of wood rises and stays above 35 percent and the temperature stays between 18° and 35°C (64° and 95°F), wood rot can become established. Serious structural damage can result.

Water, in its various forms, is probably responsible for most homeowner complaints. Common problems related to moisture include:

*   condensation on windows
*   deterioration of window frames and wall finishes under windows
*   leaks in the ceiling caused by the melting of frost accumulated in unheated attics during cold weather
*   paint peeling from exterior wood sidings
*   efflorescence, spalling of brick work and the deterioration of mortar joints
*   "dusting" or staining of ceilings and walls in exterior corners and along joists and studs or on nail or screw heads
*   mould and mildew in these same locations
*   "basement leaks" from ice melting off the foundation wall behind the insulation
*   mould and mildew on rafters, trusses or roof sheathing
*   sagging of roof sheathing
*   deterioration of the structural framework of the building due to rot
*   ice damming and the formation of icicles, and
*   water leakage from fans resulting from condensation in uninsulated ducting that runs through unheated space.

Most of these problems can be eliminated by controlling indoor humidity levels through exhaust fans, and by stopping the warm moist air from entering and accumulating in the insulated cavities of the building envelope.

2.4   CONTROLLING SOUND TRANSMISSION

2.4.1   Introduction

Sound is a form of energy carried by vibrations which stimulate nerves located in the ear. Sound is measured in decibels (dB); the minimum sound level a person can hear in a controlled laboratory experiment is 0 dB. Background noise usually ranges from 10 to 50 dB, normal conversation measures about 70 dB, and conversation at a party might rise to 85 dB. The pain threshold for the human ear is approximately 130 dB.

Noise at some frequencies cause a greater distress than others. For this reason, all noise regulations use ratings that combine dB values at many frequencies. The A-weighted sound level (dBA) rating matches frequency response of human hearing in a standard way, that can be measured with a meter. The dBA rating is used for regulating noise from outdoor sources (air conditioners, road traffic, etc.). Noise emissions from many appliances are rated in dBA and also in sones for household fans. For example, sound power from exhaust fans located in kitchens should not exceed 60 dBA or 3.5 sones.

Sound waves carried through the air are either *absorbed* or *reflected* when they strike a surface. Absorbed sound energy is transmitted through an object by vibration. Reflected sound will continue to reverberate in the air, creating echoes even after the source of the sound has been stopped.

2.4.2   Mechanisms of Sound Management

Noise, or unwanted sound, can cause a major concern for many home owners since it affects their ability to enjoy their home. It can originate inside the home from furnace blowers, dryers, and stereos, for example, or from people moving around, playing or working with tools. It can also originate outside, from cars, trucks, trains, neighbours, etc. A well built home is designed and built to minimize the transmission of sound, to ensure that noise does not become a problem for the occupants.

Sound vibration can be transmitted in two ways: through the air (airborne sound) and through solids — in this case, the house structure itself (structure-borne sound).

Structure-borne sound can be reduced in two ways. The first is to use building assemblies constructed with heavy materials or built-in layers to reduce the transfer of vibration. The structural components in these assemblies should be isolated from each other to eliminate sound transmission from one to the next. This can be done using insulation or an air space, although the latter will not perform as effectively. Higher density insulations are better isolators than low-density products. Soft floor coverings will also help to reduce sound transmission from one level in a building to the next.

The second approach involves the use of materials such as resilient channels to fasten finishing materials to the wooden framing members. This material helps to minimize the transmission of vibrations through the members. Resilient channels in ceilings can be very effective in reducing sound transmission from one level to the next, especially in multi-level structures.

Airborne sound can be greatly reduced. Acoustical caulking was developed specifically to seal the paths by which air leaks in or through the building. Although air tight homes are quieter, because there are fewer paths for external airborne noise to enter the home, even greater concern must be given to lessening internal noise. Weatherstripping interior doors and using air tight electrical boxes can reduce airborne noise travelling from one room to another. The construction techniques described in later sections of this manual are designed to minimize air leakage through all components of the building envelope and thus help reduce airborne sound.

However, you may need to pay special attention to internal partitions and floors if some rooms or areas of the house are to be used for noisy activities, especially in apartments and row houses.

2.4.3   Control Potential

Sound is an issue that should be dealt with at the design stage. The National Building Code of Canada specifies that all constructions separating dwelling units in multi-unit buildings must have a minimum Sound Transmission Class (STC) of 50. The higher the STC rating of a system, the greater the reduction in sound transmission. An STC of 55 represents a better design goal. To minimize the transmission of sound into a house, and from unit to unit in a multi-unit building, follow the guidelines given below.

Site

Houses should be situated so that natural terrain, trees and other buildings provide shelter from noise generated elsewhere — from the air conditioners and heat pumps of other houses, for example, or from nearby roads or other transportation corridors.

At very noisy sites, wall and window constructions with suitable noise reduction ratings are required. Selection procedures have been developed by CMHC, NRC, and provincial regulators.

Layout

The house's floor plan should be laid out to separate noisy areas from quiet ones. Bathrooms, kitchens, closets and hallways should be used to buffer bedrooms from areas where noise will be generated, such as family rooms.

In multi-unit buildings, it is especially important to keep noise from one unit from disrupting the occupants of the next. If possible, windows and doors should not face areas which generate a lot of noise.

Mechanical Equipment

Some of the mechanical equipment installed in houses can be noisy. If possible, furnaces, clothes dryers and ventilation systems should be located in closed rooms away from sleeping areas. Proper ductwork design can assist in reducing the noise associated with forced-air furnaces. For example, vibration collars can be used to reduce the transmission of sound from the furnace to the ductwork.

Plumbing systems can also be a source of noise in housing. Air locks in pipes, water pumps and sump pumps located in basements can also be problematic. Pump noise can be eliminated by using submersible pumps.

Most noise from plumbing can be controlled by avoiding contact between pipes and wall surfaces, using extra gypsum board for walls or ceilings near pipes, and using resilient lining such as foam pipe insulation wherever possible.

Building Materials and Assemblies

Sound absorbing materials and sound insulation can be used to keep outside noise from penetrating the building, and to keep sound waves from travelling from one room or area of the house to another. To minimize sound transmission, choose wall, window, ceiling and floor materials with high STC values. Ordinary furnishings, drapes and carpets will also help to "deaden" sound.

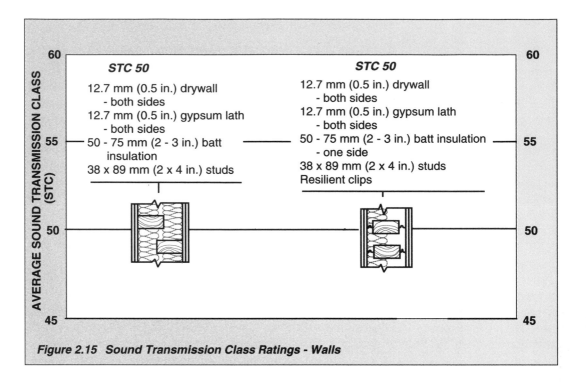

Figure 2.15 Sound Transmission Class Ratings - Walls

The wall diagrams include:

AVERAGE SOUND TRANSMISSION CLASS (STC)

STC 50

12.7 mm (0.5 in.) drywall
  - both sides
12.7 mm (0.5 in.) gypsum lath
  - both sides
50 - 75 mm (2 - 3 in.) batt
  insulation
38 x 89 mm (2 x 4 in.) studs

STC 50

12.7 mm (0.5 in.) drywall
  - both sides
12.7 mm (0.5 in.) gypsum lath
  - both sides
50 - 75 mm (2 - 3 in.) batt insulation
  - one side
38 x 89 mm (2 x 4 in.) studs
Resilient clips

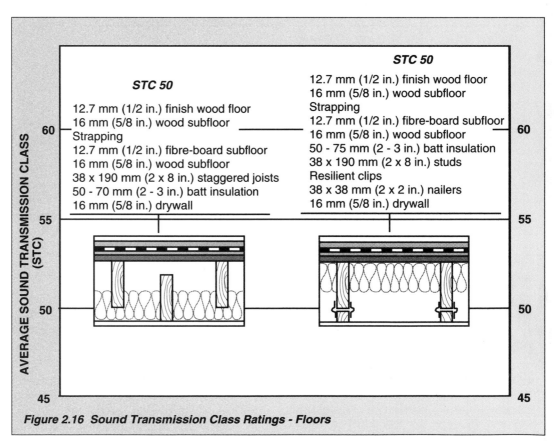

Figure 2.16 Sound Transmission Class Ratings - Floors

AVERAGE SOUND TRANSMISSION CLASS (STC)

STC 50

12.7 mm (1/2 in.) finish wood floor
16 mm (5/8 in.) wood subfloor
Strapping
12.7 mm (1/2 in.) fibre-board subfloor
16 mm (5/8 in.) wood subfloor
38 x 190 mm (2 x 8 in.) staggered joists
50 - 70 mm (2 - 3 in.) batt insulation
16 mm (5/8 in.) drywall

STC 50

12.7 mm (1/2 in.) finish wood floor
16 mm (5/8 in.) wood subfloor
Strapping
12.7 mm (1/2 in.) fibre-board subfloor
16 mm (5/8 in.) wood subfloor
50 - 75 mm (2 - 3 in.) batt insulation
38 x 190 mm (2 x 8 in.) studs
Resilient clips
38 x 38 mm (2 x 2 in.) nailers
16 mm (5/8 in.) drywall

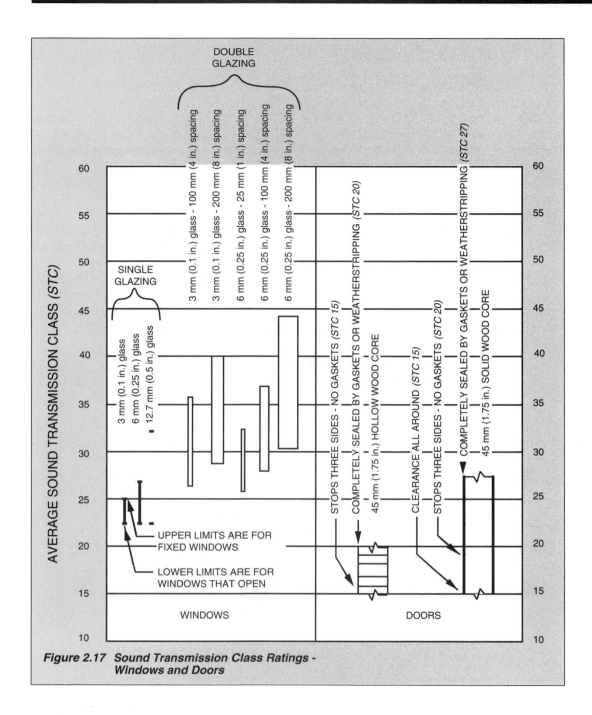

**Figure 2.17 Sound Transmission Class Ratings -
Windows and Doors**

Figures 2.15, 2.16 and 2.17 illustrate some typical wall, floor, window and door assemblies, and provide average STC ratings for each. Windows are clearly the biggest problem in terms of sound transmission, followed closely by doors.

Sources for Further Information

CMHC has developed several publications to assist the residential construction industry. Many of the gypsum wall board and insulation manufacturers have also prepared materials to minimize noise problems associated with residential construction.

2.5 OTHER FACTORS AFFECTING COMFORT

Feeling comfortable depends on more than just air temperature. It also depends on the presence or absence of drafts, the temperature of surrounding surfaces, and the temperature of the surface with which one is in contact. A warm surface radiates heat to a cooler surface. The relative humidity level and air quality will also greatly affect comfort and health.

2.5.1 Relative Humidity

Expressed as a percentage, the term "relative humidity" refers to the quantity of water vapour in the air relative to the amount of water vapour it could hold at a given temperature.

The relative humidity can range from zero to 100 percent. Above 100 percent (the dew point), the air mass can hold no more water vapour and condensation occurs in the form of dew, frost, rain, ice or snow.

Human beings are comfortable within the range of 20 to 85 percent relative humidity, but studies show the optimum range is between 30 and 55 percent (see Figure 2.18). Inadequate humidity can cause static electricity and dry, scratchy throats. Excessive humidity can lead to mould growth and excessive condensation on cool surfaces such as windows. For health and comfort reasons, however, consumers often want humidity that cannot be tolerated in our housing without causing severe frosting/condensation on windows. These problems in turn lead to home owner complaints.

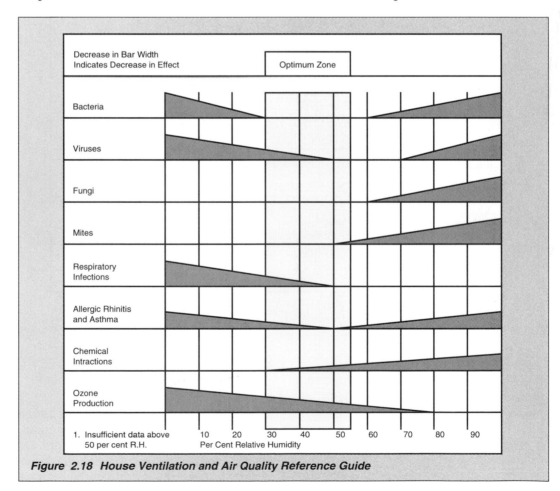

Figure 2.18 House Ventilation and Air Quality Reference Guide

2.5.2   Air Temperature

The human comfort zone in air temperature ranges from 20°C (68°F) to 28°C (82°F). The "core temperature" of the human body is 37°C (98.6°F). The difference between the air temperature in a house and the core temperature of the occupants creates a continual heat exchange.

The rate at which this exchange occurs is critical to the sensation of comfort. If it is too fast, the occupants feel chilled. Conversely, if it is too slow, they feel overheated. Individuals at rest or doing "desk work" will require higher air temperatures to feel comfortable than those who are more active or performing physical work.

2.5.3   Radiant Exchange

The occupants of a house will also radiate heat to cooler surrounding surfaces such as uninsulated walls or windows. Surfaces which are warmer than their bodies will transfer heat to them by radiation. The rate of this heat transfer depends on the difference in temperature between the surfaces and the occupants, and on the rate at which each emits radiant heat (emissivity). The slower the rate of heat transfer, the more comfortable the environment will feel.

For example, the occupants of a house will be warmed by sunlight radiating through windows, but will lose heat to walls and windows which are cooler than body temperature. High levels of insulation in the walls and ceilings of a house will keep these surfaces warmer in winter and make the occupants feel more comfortable.

2.5.4   Air Movement

Air movement across the skin (convection) will increase heat loss, making the occupants of a house feel cooler.

The design of heating, ventilation and air conditioning (HVAC) systems must take this air movement into account.

- Air provided by the ventilation system will be relatively cool, even with heat recovery. Such air should not enter an occupied zone before it has been mixed or heated to room (ambient) temperature.

- Even heated air seems like a cool draft if it is below body temperature. This sensation is increased just after a bath or shower, for example, when moisture is evaporating from the surface of the skin. The location of warm air ducts should be carefully planned to provide the most comfort to occupants.

- Heated air will be cooled as it comes into contact with cool surfaces such as window glass. As it cools, the air becomes more dense and falls, creating drafts at the floor level. Avoid installing cold air returns in locations where cool air will be drawn across wide expanses of floor.

- Air stratification can create drafts at floor level. A strategy that provides continuous air circulation, such as the use of two-speed furnace fans, will help to prevent these floor level drafts.

More detailed information about the design and installation of HVAC systems is provided in Chapters 12 through 18.

2.6   INDOOR AIR QUALITY

Occupants of a house require fresh air to remain healthy and comfortable. Stale, moist air must therefore be removed from the living space.

For many years, this air was supplied randomly by air leakage which provided natural ventilation through cracks and holes in the building envelope. The pressures created by wind, the stack effect and the operation of appliances caused large amounts of fresh outdoor air to enter the house. However, as construction techniques and codes have changed and houses are being built with tighter envelopes, these pressures have proved inadequate to ensure sufficient ventilation.

At the same time, the use of synthetic building materials and home furnishings made with potentially harmful chemicals has increased. Over time, these materials release chemicals into the air through a process known as "outgassing". The principle behind outgassing is similar to the principles behind heat transfer and moisture movement: chemicals will move from an area of high concentration (the synthetic material) to an area of low concentration (the surrounding air). Potentially harmful chemicals in solvents and cleaners used by home owners can add to this problem.

More recently, moulds and mildew in housing have been targeted as a health hazard. The spores produced by these mould growths can circulate through the house air and act as an irritant and/or catalyst to health problems. Similarly considerable attention has been given to the potential adverse health effects of radon in areas where it occurs in the soils.

These issues highlight the problems of indoor air quality. In many cases there will be no identifiable cause of health related problems for the occupant. Source control of pollutants and contaminants and dilution with outside air must be considered essential in providing acceptable indoor air quality.

More detailed information about designing and building to provide improved indoor air quality and healthier housing is provided in Chapter 3.

Chapter *3*

DESIGN CONSIDERATIONS

The previous chapter reviewed the fundamental principles that underly the construction techniques described in this manual. This chapter looks at factors you should consider during the design process. It reviews how various design decisions arising from the application of these principles can affect the durability, comfort, energy efficiency, and overall marketability of houses.

3.1 DESIGN DECISIONS

Because of the way house systems interact, design decisions can affect the performance of the house in many ways. The design process is a series of tradeoffs, all of which affect the performance of the house, its marketability, and the comfort of the occupants.

- North-facing windows may be installed because of views, but may require different types of glazing. Additional envelope insulation may be needed to compensate for the loss of energy performance.

- An open plan may facilitate air circulation, especially desirable in homes with woodstoves or those incorporating passive solar design, but the increased noise and reduced privacy may offset some of the benefits.

- Taking advantage of passive solar gains may increase capital costs, but reduced operating costs and pleasant, sunnier rooms may make it worthwhile. Care must be taken at the design stage to ensure the increase in solar energy doesn't increase summer cooling loads.

- High levels of insulation and a well-sealed building envelope will reduce the size and cost of heating and cooling equipment required.

- Installing a heat recovery ventilator (HRV) in a tightly sealed house may increase initial costs but will result in long-term energy savings.

Plan your heating and ventilation system designs in conjunction with your house design. If you don't call in your heating contractor until after the house is already framed, you will seriously limit your options. Also consider the type of household and its requirements, the type and location of windows and vents, the availability and cost of fuels, the noise generated by the equipment, and how easily the equipment can be maintained. By considering all these factors early in the design process, at a general or sketch level, you can ensure that the house subsystems are working together, not against each other.

Of course, the best design can result in a poor product if difficulties arise during construction. Good organization can minimize such difficulties. Key things to consider include scheduling, the presence of the right skills at the right time, the availability of materials and equipment, and weather conditions during critical periods, especially before closing in.

Scheduling problems can be severe in remote and northern areas where the construction season is very short and skilled local workers are often not available. The fact that materials must often be shipped in by winter road, barge or air adds to costs and affects timing. Ensuring that the required materials are available when needed demands considerable preplanning and preordering, close supervision of shipping, careful storage, and tight control on the site.

3.2 SITING AND LOT PLANNING

The first step in design is to select the site. It must be readily accessible to vehicles and pedestrians, and serviceable. For purposes of energy efficiency, lots should have good solar exposure. Individual lots in subdivisions should be laid out carefully so that each is provided with the maximum available solar exposure (see Figure 3.1). This exposure is enhanced by orienting the majority of streets on an east-west axis.

Remember that heat from the sun is only desirable during the heating season.

- Locate trees and shrubs to provide summer shade to east, west, and south-facing glazing.

- Deciduous trees provide good solar control in northern climates but let sunlight through during the winter (see Figure 3.2).

- Landscape with coniferous trees to provide shelter from prevailing winds. In windy areas, good wind breaks can reduce air leakage from a house by 30 to 40 percent by reducing wind pressures.

Northern locations present their own challenges. Soil bearing capacity can be a major problem where conditions include frost-susceptible soils, muskeg and permafrost, and where deep frost penetration makes it difficult to install services. Study local practice and consult experienced professionals when choosing sites.

- Site to gain sunlight in spring.

- Snow drifting can be a major problem. Try to determine the direction of the prevailing winter wind and orient the long axis of the building in the same direction.

Whenever you build a house, take advantage of the experience of builders and residents in the area. You may find there are good reasons for practices that may seem at first to be odd or unusual.

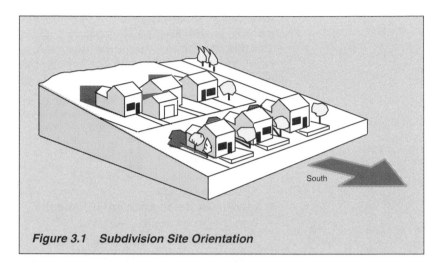

Figure 3.1   Subdivision Site Orientation

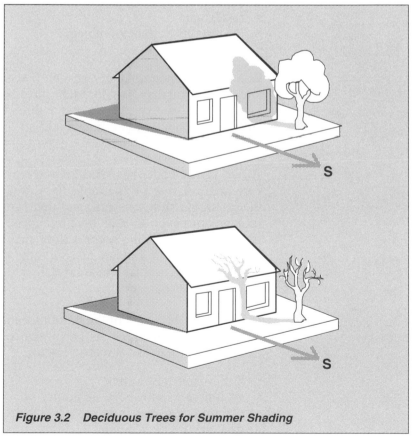

Figure 3.2   Deciduous Trees for Summer Shading

3.3   BUILDING DESIGN

3.3.1   Configuration

The first step in designing the house itself is to consider its overall configuration and its location and orientation on the building lot. It is not always immediately clear whether the house should be one storey, two storeys or split level, or whether a basement is desirable. In making these decisions, consider the following factors:

- site and climatic conditions
- the neighbourhood
- client needs and requirements
- functional usability of spaces and components
- quality of the interior environment
- energy performance, verified through the use of computer programs such as HOT 2000®
- capital and operating costs
- resale, and
- the overall character/ambience of the home.

Look at the topography of the site. If the site slopes, you may be able to take advantage of the slope to produce a two or three-storey configuration at one end and a reduced profile at the other.

A two-storey house has less exposed surface area relative to its usable floor area than a single-storey design, and this can reduce construction and heating costs. Better views from the second floor may also be a factor. A measure of a building's efficiency is its usable floor area relative to its exposed surface area.

In the north, the shape of the building can have a major impact on its performance because of the severe climate. Try to:

- minimize surface area
- minimize windows facing north
- place entries away from snowdrift locations, and
- locate unheated entry vestibules so they project out from the building.

Basements are seen as necessary in many areas by consumers, the assumption being that they are inexpensive and necessary for resale. As building codes and consumer expectations change, this may not always be the case.

Basements or ground floor work areas are especially useful in remote and northern areas where there is often a greater need for storage, workbenches and repair activities. However, *crawlspaces* may be more practical because of the deeper frost penetration in northern regions, or high water tables in some areas. In permafrost areas, basements are possible only in unusual circumstances, and open crawlspaces are common. In southern areas, bi-level designs or units with basements are usually cost-effective since they add considerable space at a marginal cost.

The *type of roof* is another important consideration. The conventional approach is to construct a pitched roof with prefabricated high-heel or drop-chord trusses, but there are other good alternatives which allow for increased insulation and reduced lumber requirements.

- Scissor trusses or truss rafters can be used to build cathedral ceilings.

- If using cathedral ceilings, give careful consideration to the relationship of interior partitions to the ceiling, and to the effect of an open plan on the occupants' lifestyle and the operation of mechanical systems.

- In the north, truss rafters are being used in unventilated roofs to reduce problems associated with snow infiltration.

Solar energy which typically accounts for 15 percent of an average home's heating can provide up to 40 percent in a well designed house, without compromising lifestyle. A properly designed house takes advantage of passive solar heating without overheating or glare.

Design considerations include: good orientation; proper window selection, particularly important with the new glazings; suitable floor plans; appropriate shading; an increase in house mass; and a distribution system to circulate heat.

- To obtain maximum benefits from solar energy, about 50 to 60 percent of the total window area should face within 30 degrees of due south (see Figure 3.3). Requirements for ventilation or views may temper these guidelines. Since our homes are typically designed for curb appeal, this may require adjustments to consumer expectations.

- The area of non-south-facing windows should be reduced to the minimum required by building codes or as close to that as the purchaser will accept.

- Windows should be selected using the information provided by manufacturers testing to Canadian Standards Association (CSA) Standard A-440.2. It is now common to find fixed windows that perform better than an R-20 wall — providing a net heat gain over the heating season.

- Sloped glass and skylights should be triple-glazed and equipped with low-E (low-emissivity) coatings (see Chapter 10).

The orientation, the window area, and the ability of the glass to transfer heat (the solar heat gain coefficient) determine how much solar heat enters the home. "Thermal mass" refers to a building's ability to store or retain solar heat overnight and on days when there is little solar energy input.

A standard wood-frame house with conventional windows is built with enough drywall, floor coverings, wood, etc. to provide sufficient thermal mass for a south-facing glazed area to equal 6 to 9 percent of the total floor area.

This amount of south-facing glazing will contribute significant solar heating without overheating (see Figure 3.4). Exceeding that percentage will require increased mass to absorb the solar heat and/or proper window selection to minimize problems with overheating and also chilling at night. Remember, too much heat can be as big a problem as too little.

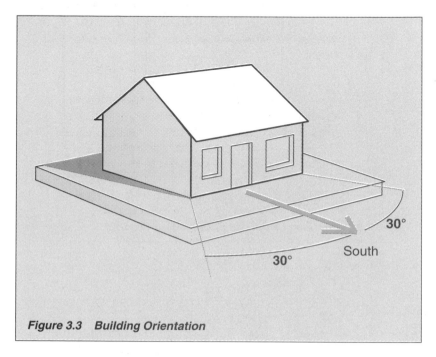

Figure 3.3   Building Orientation

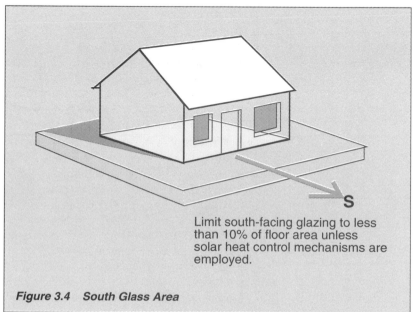

Limit south-facing glazing to less than 10% of floor area unless solar heat control mechanisms are employed.

Figure 3.4   South Glass Area

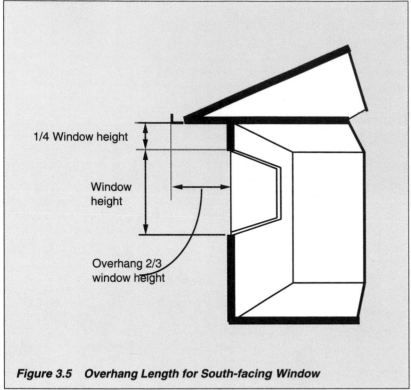

1/4 Window height

Window height

Overhang 2/3 window height

Figure 3.5   Overhang Length for South-facing Window

You must also provide some means of controlling the entry of solar energy.

- Select a glass that matches the needs of the design for visible light, solar gain and R-value.

- Complement the window selection by designing overhangs above the windows to shade the glass when solar heat is not required (see Figure 3.5).

- South-facing windows are relatively easy to protect, but east- and especially west-facing windows are more difficult to deal with because of the lower sun angle during the summer (see Figure 3.6).

- Use exterior louvres, screens, latticework and awnings.

- Consider internal window shading devices if necessary, but keep in mind that these do not reduce overheating as effectively because the solar radiation enters the building before it is blocked.

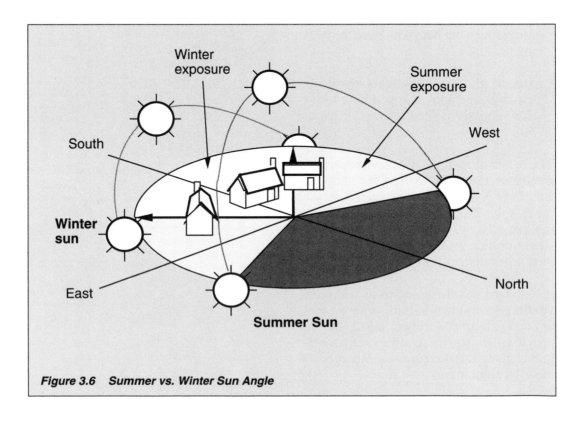

Figure 3.6   Summer vs. Winter Sun Angle

3.3.2 Layout and Floor Planning

A functional floor plan has several basic features:

- the efficient use of space in each room — use scale drawings of furniture to assess the proposed layout
- a minimal number of circulation paths which reduce useful space
- a proper progression from areas which are "public" to those which are "private", called the "intimacy gradient"
- a plan to reduce interior noise
- the location of hallways, utility rooms, bathrooms, etc. on the north side of the home, and
- the appropriate use of passive solar gains and daylighting.

Refer customers to their existing house or apartment to determine whether rooms are large enough and laid out comfortably.

The new design should improve on the areas they feel are inadequate in their current home.

Beyond the purely functional requirements, consider aesthetic and comfort factors. The layout should exploit the best views. Rooms such as the living and family areas should be located on the south side of the building to get the most benefit from passive solar gains and daylighting (see Figure 3.7). The layout of rooms and spaces is especially important in northern locations where winters are long and dark and occupants spend most of their time indoors.

Recent CMHC studies have determined a willingness on the part of consumers to trade off overall house size, in return for more compact, better designed, and better appointed spaces.

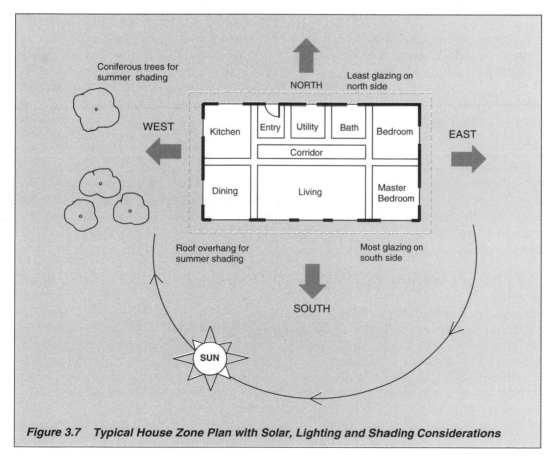

Figure 3.7 Typical House Zone Plan with Solar, Lighting and Shading Considerations

3.3.3 Natural Ventilation

Where possible, the house should be oriented, and windows located and sized, to take advantage of natural ventilation from prevailing winds. The effectiveness of natural ventilation schemes also depends on the layout of partition walls, and on the location of corresponding windows and doors for cross ventilation (see Figure 3.8).

3.3.4 Lighting

Good house design allows for high levels of natural lighting, particularly in areas which are normally occupied during the daytime. Keep these points in mind during the design process.

- In areas of high use such as kitchens, bathrooms and central hallways, plan for the use of efficient lighting such as compact fluorescent, high efficiency fluorescent and halogen lights.

- Kitchens, breakfast areas, and living rooms should be located to receive as much daylight as possible.

- High windows, skylights and "light pipes" can provide the most natural light to a given area.

- White or light-coloured interior finishes will enhance the effects of daylighting.

Figure 3.9 illustrates several design strategies you can use to take advantage of natural lighting.

3.3.5 Privacy and Security

Privacy and security are two other factors that should be taken into consideration when designing a house. Privacy can be maximized by orienting the house on the lot to shield yards and decks from the street and other houses as much as possible, and through the judicious use of trees, shrubs, hedges, and fences.

Security has become an increasingly important issue in recent years. Select windows and doors with hardware that cannot be easily forced open, now a building code requirement in several jurisdictions. Patio doors should be equipped with a bar that prevents the door from opening when in place. Basement windows can be protected with bars. Some insurance companies require such measures in urban areas. Exterior doors should be located in a manner that makes them visible to the street or to adjacent homes. Doors and frames should be designed to prevent forced entry, and be fitted with glazed panels or viewing holes to permit occupants to determine who is at the door before they open it.

Use solid core exterior doors equipped with deadbolts with 25 mm (1 in.) minimum throw. Many wood panel doors do not provide acceptable security. Door hinges should have screws at least 25 mm (1 in.) long going into the door, and 31 mm (1-1/4 in.) going into the frame. The framing around door openings should include solid blocking at lock height on both sides of the opening. Walls around doors should be covered with plywood or waferboard sheathing to prevent "jamb spreading" in which prybars are pressed against the door to increase the opening and render the lock ineffective.

Exterior lighting can also be used to help make the house more secure. Use fixtures with switches rather than the "dusk-to-dawn" variety so that the owners can shut the lights off to reduce electricity consumption. Where possible, use energy efficient lighting. Contact your electrical utility for the most up-to-date information.

In many cases, you may want to prewire an alarm system. There are several types available, operating on different principles. For best results, consult a specialist in the design and installation of these systems.

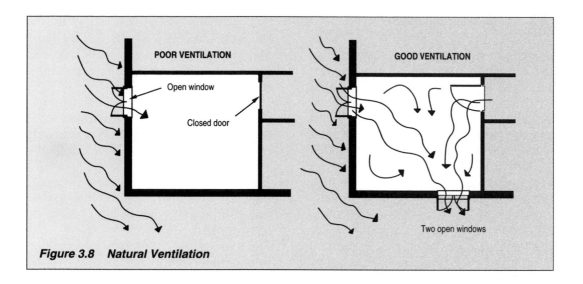

Figure 3.8   Natural Ventilation

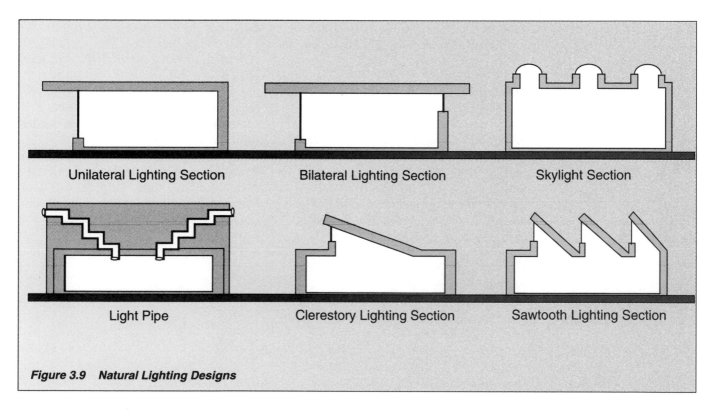

Unilateral Lighting Section

Bilateral Lighting Section

Skylight Section

Light Pipe

Clerestory Lighting Section

Sawtooth Lighting Section

Figure 3.9   Natural Lighting Designs

3.3.6 Material Selection

An increase in allergies, asthma, and chemical sensitivities has created a growing number of buyers who are concerned about the materials used to construct and furnish their homes. This concern relates directly to an overall concern for the environment. To date we have assumed that mechanical ventilation in tighter houses can provide the solution for indoor air quality for most consumers but, depending on a person's sensitivity, field studies now point out that source control or proper material selection may be even more important in the long run. This is a new and highly specialized field. To assist builders and homeowners in this task, CMHC is preparing both a booklet and computer disk with valuable information on materials and their suitability.

R-2000 Indoor Air Quality Requirements now reference:

- Carpeting; carpeting (except as noted) shall cover no more than 50% of the interior floor area. Carpeting not containing glues, with no underpad or with low toxicity underpad are exempt. The interior floor area includes the basement floor area. (the following floor coverings are exempt: wool or carpet area rugs and carpeting that has latex free backing).

- Air filtration; an air filter with a minimum 10% ASHRAE average dust spot efficiency shall be installed where air circulating heating, cooling and ventilation systems are used.
- Paints and varnishes; all liquid coatings used indoors except on wood floors, are to be water based, interior type or meet or exceed Environmental Choice standards. Prefinished items are allowed.
- Flooring Adhesives; all finish flooring adhesives to be either water dispersion or pre-adhesive types or have low toxicity content.
- Kitchen cabinets and bathroom vanities; cabinets and vanities made from manufactured wood products shall be made from formaldehyde free particle board, or particle board meeting the E-1 European standard, the U.S. HUD standard, or have all exposed surfaces sealed or a low toxicity content sealer.
- Wood flooring; all liquid coatings used on wood flooring shall meet or exceed Environment Choice standards or be prefinished.
- "Exterior grade" wood composition boards and plywoods that use phenol-formaldehyde glues in place of urea formaldehyde glues for use inside the home.

3.4 ADDITIONAL DESIGN CONSIDERATIONS FOR MULTI-UNIT BUILDINGS

Multiple-unit buildings tend to lose considerable heat through thermal bridges. Many of these thermal bridges can be avoided, or their effects minimized, if they are recognized at the design stage.

- Masonry, steel and concrete conduct heat readily. Design horizontal fire separations as firewalls only if required by code. Firewalls expose more surface area to the outdoor environment (see Figure 3.10). Also Some insulations cannot be used in firewalls because of combustibility.

- Avoid thermal bridges such as cantilevered slabs which extend beyond the exterior walls to form balconies.

- Design party walls to minimize heat loss and air leakage (see Section 8.14).

Many consultants and architects have now adopted the use of computer simulation as a means to check, at a design stage, decisions being considered. These sophisticated computer software programs can help identify the best design options for the building envelope and the heating, cooling, and ventilation system selection.

When selecting windows pay special attention to window size, air leakage and glazing type. New glazings can control both heat loss and heat gain. The use of information provided by the performance testing of windows to CSA Standard A-440 is helpful when selecting windows. Properly manufactured and installed windows will lessen the potential for discomfort due to overheating, drafts, or the "cold wall effect" i.e. the excessive radiation from an occupant's body to the cold window surface. They may also reduce the need for additional framing members in the walls.

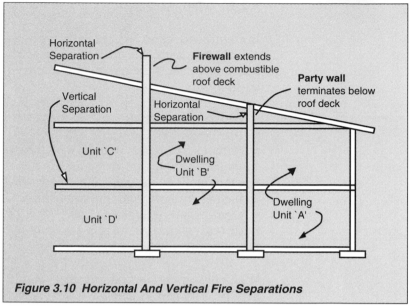

Figure 3.10 Horizontal And Vertical Fire Separations

3.4.1 Sound Transmission

Carefully consider floor plan layouts including the locations of windows and doors. Provide buffer zones between active, noisy areas such as living rooms and quiet zones such as bedrooms, and locate windows and exhaust vents to prevent noise flanking (see Figure 3.11) from one unit to another. In multi-level buildings, locate noisy areas in one unit away from quiet areas in the unit above or below.

All construction assemblies separating one unit from another should have sound transmission classification (STC) ratings of 50 dB (see Section 2.4.3).

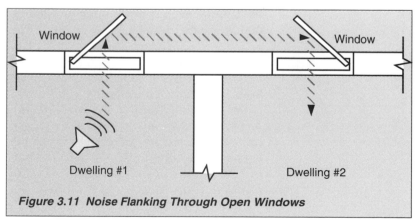

Figure 3.11 Noise Flanking Through Open Windows

3.4.2 Airtightness

In addition to the air leakage paths found in detached houses, the following leakage paths have been noted in multiple-unit buildings:

- lack of firestopping
- furring chases
- unblocked cavities in hollow masonry walls
- electrical penetrations and other service penetrations through party walls
- stairs adjacent to party walls
- doorways into common areas, and
- adjoining pipe chases.

When designing the building, you should give priority to designs and strategies that will reduce air leakage from these areas as much as possible. Air leakage greatly affects energy performance, durability of the structure, occupant comfort, and noise infiltration.

3.4.3 Fire Separation

In a residential building where no dwelling unit is located above another, for example, townhouses, a party wall on a property line need only be a fire separation with a one hour fire-resistance rating. It does not have to be a firewall. Continuous protection must be provided from the top of the footings to the underside of the roof sheathing. If a gap is left between the top of the party wall and the roof sheathing, it must be tightly sealed with noncombustible material (see Figure 3.12).

The assemblies described in Figure 3.13 have a fire resistance rating of about one hour, as calculated using the *Supplement to the National Building Code of Canada*. Actual ratings can be determined from tests conducted by a recognized testing agency according to CAN4-5101, *Standard Methods of Fire Endurance Tests of Building Construction and Materials*.

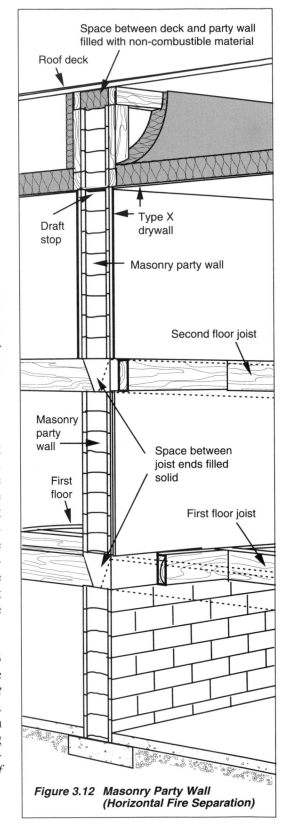

Figure 3.12 Masonry Party Wall (Horizontal Fire Separation)

	No.	Assembly	Finishes (*) (**) (***)
Walls	1)	150 mm (6 in.) poured concrete wall.	Both sides sealed with two coats of paint.
	2)	140 mm (5.5 in.) hollow concrete block wall, 25 mm (1 in.) mineral fibre insulation.	12.7 mm (1/2 in.) gypsum wallboard, resilient channels on at least one side.
	3)	Single row wood stud wall, 38 x 89 mm (2x4 in.) studs @ 400 mm (16 in.) o.c., 75 mm (3 in.) mineral fibre insulation.	Double layer of 12.7 mm (1/2 in.) gypsum wallboard each side, resilient channels each side.
	4)	Staggered wood stud wall, 38 x 89 mm (2x4 in.) studs @ 400 mm (16 in.) or 600 mm (24 in.) o.c. on common 38 x 140 mm (2x6 in.) bottom plate, 75 mm (3 in.) mineral fibre insulation.	Double layer of 12.7 mm (1/2 in.) Type X gypsum wallboard each side, resilient channels each side.
	5)	Double row wood stud wall, 38 x 89 mm (2x4 in.) studs @ 400 mm (16 in.) o.c. on separate bottom plates spaced 25 mm (1 in.) apart, 75 mm (3 in.) mineral fibre insulation.	16 mm (3/5 in.) Type X gypsum wallboard, resilient channels each side.
	6)	Single row metal stud wall, 90 mm (3-1/2 in.) studs @ 400 mm (16 in.) o.c., 75 mm (3 in.) mineral fibre insulation.	Double layer of 12.7 mm (1/2 in.) gypsum wallboard each side, resilient channels each side.
Floors	7)	130 mm (5 in.) reinforced concrete floor.	None.
	8)	Open web steel floor joists, 50 mm (2 in.) thick concrete deck.	15.9 mm (5/8 in.) Type X gypsum wallboard ceiling, furring channels @ 600 mm (24 in.) o.c.
	9)	Wood floor joists @ 400 mm (16 in.) o.c., 15.5 mm (5/8 in.) plywood or waferboard subfloor, 50 mm (2 in.) lightweight concrete topping, 75 mm (3 in.) mineral fibre insulation.	Double layer of 12.7 mm (1/2 in.) Type X gypsum wallboard ceiling, resilient channels.

Notes:

*   The designated finishes are required for fire ratings of 1 hr. If a lesser fire resistance rating is required, changes in finishes are possible.

**   Sound resistance of wall assemblies can be improved by increasing the thickness of the wallboard finishes indicated. However, four (4) layers of wallboard appear to be the optimal limit.

*** As indicated, floor finishes have not been included. To assist in reducing impact noise, carpets and floating floors can be considered.

Figure 3.13   *Examples of Building Assemblies Having at Least a 1 Hour Fire Resistant Rating and STC of 50 or More*

3.4.4 Ventilation

The possibility of cross-contamination between exhaust air and incoming fresh air is greater in multiple-unit buildings. In addition to the requirements for detached houses (see Chapter 16), the following factors should be considered.

- The envelope area available for placement of exhaust and inlet ducting in a multi-unit building is restricted.

- The exhaust or intake of one unit should be located at least 2 m (6 ft.) away from the window of an adjoining unit, unless specific steps are taken to prohibit cross-contamination.

- If necessary, consider locating exhaust outlets above the roof-line of the building so that warm exhaust air will rise away from the inlets.

3.5   INDOOR AIR QUALITY

Providing good indoor air quality is part of good design practice. On average, people spend over 90 percent of their time indoors.

The quality of indoor air is important to the occupants' comfort and health. Increasingly, people are affected by health conditions such as allergies and asthma which are related to indoor air quality. For these people, their home should be an oasis where they can get respite from their symptoms.

Indoor air quality is dependent on many factors.

- The building site determines the ambient air quality and whether radon is a factor.

- Many building materials add chemicals to the air through outgassing.

- Selection of heating equipment can contribute to particulates and noxious fumes in the air stream.

- Construction details for basement damp-proofing can affect the amount of moisture in the air.

- Occupant activities such as smoking, washing, housekeeping, hobbies and even operation of the ventilation system can seriously affect indoor air quality.

The designer and builder cannot affect ambient air quality or occupant activities but can incorporate the three indoor air strategies — eliminate, separate, and ventilate — into various aspects of the house design. The builder can design the building so that its materials and construction promote good indoor air quality.

The builder can include an appropriate ventilation system as part of the design, and even mitigate effects of ambient air quality through special ventilation strategies. Finally, the builder can promote awareness of indoor air quality and occupant activities to new owners through providing an owners' manual with the house, for example.

The design strategies of good indoor air quality and R-2000, or super-insulated houses, are complementary. Both look to a well-sealed building envelope, controlled ventilation and low heating requirements. Ventilation strategies related to indoor air quality are dealt with in Chapter 16.

There are many additional design strategies which can be used to improve indoor air quality. This section will look at design issues related to indoor air quality at several steps in the design process, through:

- floor planning and layout
- the choice of materials and materials handling
- equipment selection, and
- construction details.

Each section discusses good standard practice and "extras" which can improve indoor air quality for the average homeowner. At the same time, some responsibility for indoor air quality falls within the homeowner's realm. Homeowners must understand the importance of operating ventilation systems and they must understand that they are responsible for many of the toxic materials (cleaning agents, hobby materials, etc.) entering the house.

Special publications are available which detail the measures required for individuals with extreme health problems related to indoor air quality. These are listed in the Appendix.

3.5.1 Layout Design Strategies

The quality of indoor air which is enjoyed or suffered by the homeowner is affected by numerous decisions made throughout the design and construction process. Beginning with the site plan, main living areas should be located to take advantage of prevailing summer breezes for cross ventilation and natural cooling. The site plan should also carefully consider the location of the ventilation intake air; this intake should be away from busy roadways, driveways or uses such as a neighbour's garage or a commercial property.

While developing the floor plan, you have an opportunity to separate the garage from the house entirely. If this is not possible, the garage should be carefully sealed from the house and separate exhaust ventilation provided for the garage.

The house layout stage also provides an opportunity to group rooms into air quality "zones". Basically, the bedroom area should be the "cleanest" area of the house and well away from pollutant sources such as the kitchen, laundry room, home office or hobby room. In planning the layout, it should be remembered that the basement is inside the building envelope and is therefore part of the living area as far as indoor air quality is concerned. The basement should be moisture-proofed, heated and ventilated accordingly.

It is customary in the building field to develop the design first and then "fit" the mechanical system into it. However savings and performance improvements can be obtained if heating and ventilation systems are considered from the outset. For example, grouping the laundry room, kitchen and bathrooms in the same zone will allow for shorter exhaust air ducting. Exhaust requirements for other areas such as hobby rooms, home offices or closets should also be considered when locating these rooms during the design process.

3.5.2 Materials Selection and Handling

In terms of materials selection, the materials which have the greatest implications for indoor air quality are wood products and flooring, especially carpeting.

The main problem associated with composite wood products is the outgassing of formaldehyde used as a binder in the materials. Two types of resins are used as binders: urea-formaldehyde resins used in interior grade plywood and particleboard, and phenol-formaldehyde used in waferboard and exterior plywood. Because of the resins' susceptibility to moisture, the formaldehyde emission rate is higher from interior plywood and particleboard than from exterior plywood or waferboard.

One strategy to deal with this problem is to reduce or *eliminate* the source by substituting materials, such as exterior grade plywood, which have lower formaldehyde emission rates meeting the E-1 European standard or the HUD standard, 24 CFP Part 3280.308, or materials which don't contain formaldehyde at all such as dimensioned lumber (see Figure 3.14a). Another strategy is to *separate* the source by sealing the material at its edges before installation with a product which will prevent most of the outgassing from occurring.

Foundations, framing and sheathing are outside the air barrier and are less problematic than materials which are inside the shell of the house. However, designs for extremely sensitive clients typically avoid the use of pressure treated wood for foundations and landscaping, and the use of plywood or waferboard as exterior sheathing.

There are two main problems associated with modern flooring materials: outgassing from synthetic materials, and the entrapment of dirt, mould and other contaminants.

For most materials, outgassing is greatest for a short period after manufacturing, then continues at a lower rate over an extended period of time. The effects of outgassing can be lessened by allowing new materials to "cure" before installing them in the house. Substituting sheet linoleum for tile reduces outgassing by reducing the amount of glue required for installation. Substituting natural for synthetic carpeting materials can be very helpful.

Fibrous carpeting readily traps dust particles and biological materials such as pet dander. If the carpet cannot be lifted, as in wall-to-wall installations there is a permanent build-up in the carpet which cannot be removed, even by regular vacuuming and shampooing. In addition, carpet which is installed in areas susceptible to moisture, such as bathrooms, kitchens and basements, can readily harbour mould and mildew.

Moisture related problems can be eliminated by not installing carpet in high moisture areas of the house. However, the problem of trapping dirt, particles and biological contaminants can only be addressed by substituting smooth flooring for wall-to-wall carpet and by using area rugs which can be regularly lifted and throroughly cleaned (see Figure 3.14b).

Wood Products	Standard	Better	Best
Subfloor	Interior plywood	Exterior plywood or waferboard	Dimensioned lumber, low odour softwoods
Cabinetry	Particleboard	Sealed particleboard	Dimensioned lumber

Figure 3.14a   Alternative Wood Products for Indoor Air Quality

Flooring and Carpeting	Standard	Better	Best
Kitchen and bath	Vinyl tile	Sheet linoleum; allow glues to outgas before installation	Ceramic tile
Living areas	Synthetic wall-to-wall carpeting	Natural fibre carpet; allow carpet to outgas before installation	Hardwood or ceramic tile with removable area rugs

Figure 3.14b   Alternative Flooring and Carpeting for Indoor Air Quality

Other Materials

Composite wood products and carpeting have been identified as the two major groups of materials affecting indoor air quality. Improvements in these two areas will significantly affect the indoor air quality in the home. However, other measures may be taken to further improve indoor air quality.

In general terms, substitute smooth surfaces such as plaster or drywall for rough and fibrous surfaces which can trap dust and are difficult to clean, for example, textured ceilings, ceiling tile, or textured plaster walls. Substitute natural fibres and papers, and dimensioned lumber products for materials with a high proportion of synthetics such as vinyl wallcoverings, composite wood panelling, plastic base-board and moulding. For clients who have extreme sensitivities, special measures will be required including the individual testing of materials as part of the selection process.

Many indoor air quality problems can be lessened by following some simple procedures on site. For example, store synthetic materials for as long as possible before installation to allow for the initial high level of outgassing. Also, allow for accelerated ventilation of the house during and after the use or installation of such products as caulking, paints and carpet to hasten the outgassing process.

For methods of reducing the occupants' exposure to the most prevalent indoor air pollutants, see Figures 3.14a and b.

3.5.3 Equipment Selection

The major indoor air quality problem associated with heating and cooling is the introduction of combustion by-products into the air stream.

Two types of by-products are of concern. Carbon monoxide is the product of incomplete combustion and is addressed by following standard safety procedures in installing and operating combustion appliances. Nitrogen dioxide is a product of normal combustion and is introduced into the indoor air stream through unvented gas stoves and unvented heaters. Significant amounts can be released into the air simply through the operation of the pilot light on gas appliances.

The first level of improvement is to use appliances with electronic ignition and to ensure that the appliance is properly vented to the outdoors. For gas-fired kitchen stoves, this means using an exhaust fan all the time the stove is operating. The next level of improvement would be a move to a sealed combustion unit, or to avoid combustion appliances altogether and switch to an alternate heating source.

A secondary problem associated with heating systems is the accumulation of dust, both on heating coils and in the ductwork. Heating systems should be designed to minimize complicated duct runs and allow for regular cleaning.

Because houses which are built to R-2000 standards or better have a lower heating load, they offer the opportunity to minimize the effects of combustion appliances by combining functions such as space and water heating (see Chapter 18).

3.5.4 Construction Detailing

Construction detailing is another area where R-2000 standards are complementary to the goal of indoor air quality. The three areas of detailing which have special implications for air quality are moisture protection, protection from fumes associated with attached garages, and radon protection in certain geographic locations.

Moisture protection and maintaining moderate house humidity levels are essential to prevent the growth of moulds and other micro-organisms. This prevention can be achieved by following good building practices such as dampproofing the foundation and providing for adequate perimeter drainage. Isolation of the living area from the garage can be achieved through attention to the air barrier detailing. To reduce exposure to radon, increase ventilation and seal cracks and openings in the basement or crawlspace floor and walls.

Chapter **4**

AIR, WEATHER AND MOISTURE BARRIERS AND VAPOUR DIFFUSION RETARDERS

Chapter 3 outlined many of the factors that you should keep in mind during the general design process. To develop the final plans for a building and to schedule construction, you need to determine the exact materials and construction methods to be used.

Houses are subjected to a range of weather conditions. Building assemblies must include elements to protect the structural members from exposure to excessive humidity levels and to ensure that wind does not adversely affect the performance of thermal insulation materials. This chapter defines four such elements — air barriers, vapour diffusion retarders, weather barriers and moisture barriers — and provides guidelines to help you ensure that structures will be durable and will perform as expected over the service life of the building. Figure 4.1 shows the symbols that will be used in diagrams throughout this manual to illustrate these key elements of building assemblies.

KEY TO ILLUSTRATIONS:

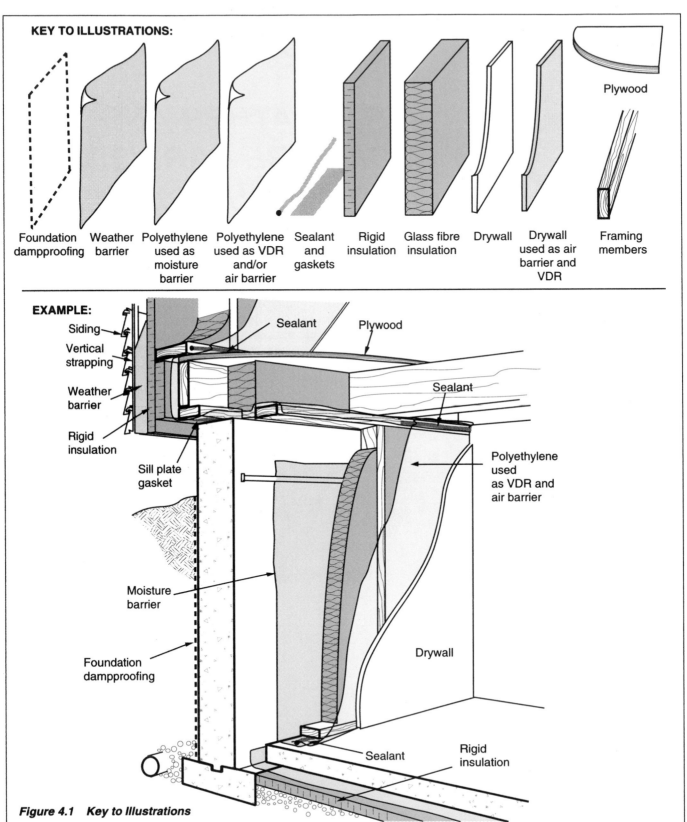

Foundation dampproofing

Weather barrier

Polyethylene used as moisture barrier

Polyethylene used as VDR and/or air barrier

Sealant and gaskets

Rigid insulation

Glass fibre insulation

Drywall

Drywall used as air barrier and VDR

Framing members

Plywood

EXAMPLE:

Siding

Vertical strapping

Weather barrier

Rigid insulation

Sill plate gasket

Moisture barrier

Foundation dampproofing

Sealant

Plywood

Sealant

Polyethylene used as VDR and air barrier

Drywall

Sealant

Rigid insulation

Figure 4.1   Key to Illustrations

4.1 AIR BARRIERS

In cold climates, air barriers are designed to reduce the outward migration of moisture-laden air (see Figure 4.2). Since almost all water vapour is carried by air movement and only a small amount diffuses through materials, the air barrier is a critical element of all components of the building envelope. The air barrier must be continuous at all corners, partition walls, floors and ceiling/wall junctions.

An effective air barrier is:

- impermeable to air flow
- continuous over the entire envelope
- able to withstand the forces that may act on it during and after construction, and
- durable over the expected lifetime of the building.

There are two types of air barriers: membrane and rigid. Both types are effective as long as they are installed properly.

- Membrane air barriers are made of materials such as polyethylene or foil. These membranes generally double as the vapour diffusion retarder (see Section 4.2).

- Rigid air barriers are made from drywall, plywood or other air impermeable materials which are sealed with gaskets and sealants to restrict air flow. Depending on the vapour permeability of the air barrier, oil-based or vapour retarding paints, polyethylene, or foil-backed drywall can be used as the vapour diffusion retarder.

Air barrier materials are discussed in more detail in Chapter 5.

The importance of making the air barrier continuous cannot be overemphasized: even the smallest cracks, holes or tears can greatly reduce its effectiveness. All joints and seams must be carefully sealed. Installation and sealing techniques for air barriers in different parts of the building assembly are discussed in Part 2 (see Chapters 6 to 10).

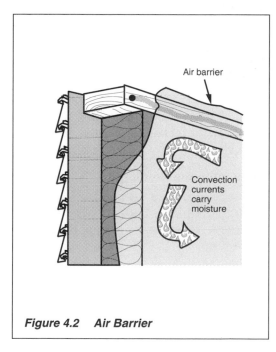

Air barrier

Convection currents carry moisture

Figure 4.2   Air Barrier

4.2 VAPOUR DIFFUSION RETARDERS

A vapour diffusion retarder (VDR) is a membrane, material, or coating that slows the diffusion of water vapour and thus helps to prevent moisture problems in houses. Polyethylene, aluminum foil and certain kinds of paint can be used as VDRs; these materials are discussed more fully in Chapter 5.

VDRs are classified according to their permeability and measured in "perms". The lower the perm rating of a material, the more effectively it will retard diffusion.

- One perm represents a transfer of one nanogram of water per square metre of material per second under a pressure difference of one pascal. In imperial units, this corresponds to one grain (0.002285 oz.) of water per square foot of material per hour under a pressure difference of one inch of mercury (1.134 ft. of water).

There are two classes of VDRs.

- A Type I VDR has a permeability of 14.735 metric perms (0.25 imperial perms) or less.

- A Type II VDR has a permeability of 43.125 metric perms (0.75 imperial perms) or less before aging and 57.5 metric perms (1.0 imperial perms) or less after aging.

Any material with a perm rating higher than 57.5 (1.0) perms is not a VDR.

Condensation will occur on any surface that is below the dew point temperature (see Section 2.3.3). The VDR can be located within the assembly as long as enough of the insulation is on the cold side to keep the assembly warm and prevent condensation. In most regions of Canada, this amounts to two-thirds of the insulating value of the insulation (see Figure 4.3). In northern areas, it may be necessary to put as much as four-fifths of the insulating value on the cold side of the VDR.

If you use the same material for both the VDR and the air barrier, it must be continuous. However, if you install a separate air barrier from the VDR, continuity of the VDR is not quite so critical, although it is still very important. If you use paint as the VDR, it must be applied evenly to the thickness required to obtain the correct perm rating.

Make sure that all materials that will be in contact with the VDR are compatible with it. If using polyethylene, ensure that it does not come into contact with heat-producing surfaces such as chimneys.

Figure 4.3 Recessed Vapour Diffusion Retarder

2/3 RSI(R)

1/3 RSI(R)

Recessed VDR

Dewpoint

Moisture laden warm air

4.3 WEATHER BARRIERS

Weather barriers protect the interior components of the wall from the effects of wind, rain, snow and sun. If wind penetrates the insulation cavity, it may reduce the effectiveness of fibre insulations such as glass fibre insulation (see Figure 4.4). If rain or snow penetrates the weather barrier, the insulation may become wet and structural members may rot.

House wrap materials have gained considerable market share over the last several years, both as sheet applied materials and as a product laminated to rigid fibreglass sheathing. In both cases, where the true potential of the material is to be realized, seams in the insulation must be sealed with a special purpose tape.

House wrap applications may not provide the airtightness desired, unless considerable attention is paid to air sealing the house interior.

For most purposes, house wrap materials should be considered as a weather barrier, not an air barrier. Where structural support is provided on both sides of the weather barrier (as discussed in 4.5.3), an effective air barrier system can be constructed.

The weather barrier is most needed where large pressure changes occur, i.e. around corners. In some northern areas, high winds can create extreme pressure changes thus creating high stresses on the material.

The weather barrier should be:

- resistant to the flow of air
- able to shed rain or snow
- able to withstand the forces that may act on it, and
- durable over the expected lifetime of the building.

Various suitable materials for weather barriers are discussed in Chapter 5.

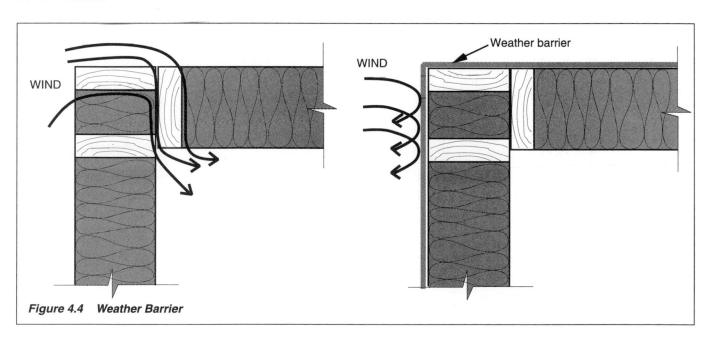

Figure 4.4   Weather Barrier

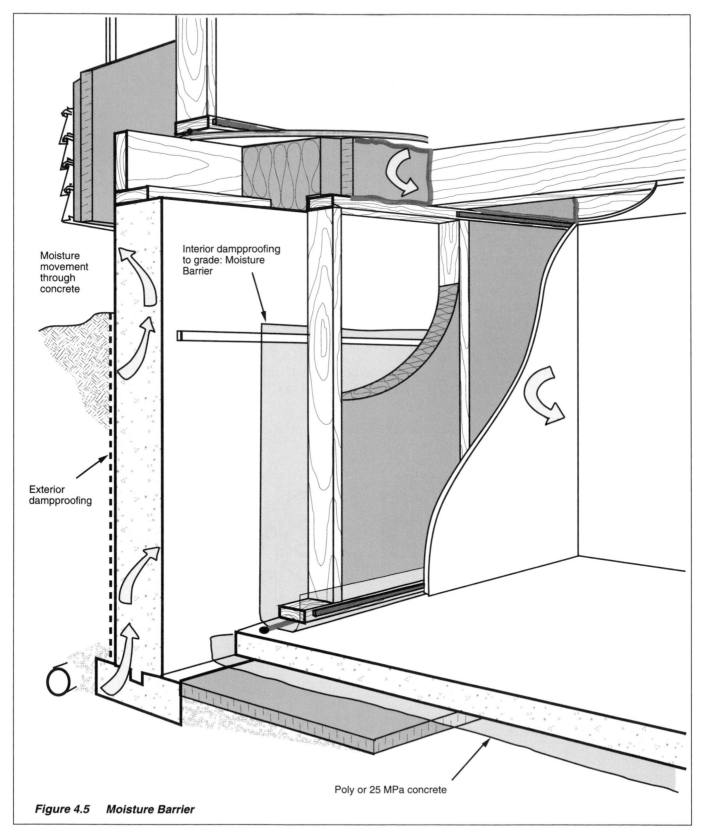

Moisture movement through concrete

Exterior dampproofing

Interior dampproofing to grade: Moisture Barrier

Poly or 25 MPa concrete

Figure 4.5   Moisture Barrier

4.4 MOISTURE BARRIERS/ DAMPPROOFING

Soil vapour pressures can cause moisture to move from the ground through the foundation wall by diffusion. Then, it is released into the air in the basement or crawlspace. To avoid this, dampproofing is required by Codes on the exterior of foundation walls located below grade. The application of a bitumen coating or polyethylene slip coat can prevent soil vapour from penetrating through the concrete.

A moisture barrier/dampproofing is also required on the interior of the foundation wall where wood and batt insulation are in contact with the concrete. A moisture barrier or dampproofing is usually provided on the interior of the foundation wall using building paper or polyethylene. One or the other is required on the below-grade portions of the foundation wall (see Figure 4.5). To ensure good coverage, the number of seams in the barrier should be minimized. Any seams should overlap and be sealed with a compatible material (see Chapter 5).

4.5 COMMON AIR BARRIER SYSTEMS

There are several different building systems commonly used to achieve a continuous **air barrier**. Most commonly. **polyethylene** is used as the primary interior air barrier in combination with sealants and header wraps; throughout this book it is referred to as the membrane air barrier approach. A second approach uses **drywall** in combination with sealants, gaskets, framing members and other rigid materials. Commonly known as a structural air barrier, or as the airtight drywall air barrier (ADA), it is referred to throughout this book as the rigid air barrier approach. Finally, **exterior air barrier systems,** represent a third option for attaining high levels of air tightness.

4.5.1 Membrane Air Barrier Approach

Polyethylene is one of the most common air barrier systems in energy-efficient houses. Its primary advantage is its dual function in providing the air barrier and the vapour diffusion retarder (VDR). To ensure its role as an effective air barrier, seams in the polyethylene should occur over solid backing, be overlapped and caulked with an acoustical caulking and sandwiched between rigid materials. Detailing at penetrations through the envelope (electrical boxes, ducting, etc.) require special attention. In many instances, an exterior air barrier (house wrap) is used to provide continuity of the air barrier between floors .

Specific detailing, sequencing and installation procedures, are provided in Part 2 of this book. Chapter 5 details specific physical properties and characteristics of suitable polyethylene materials and compatible sealants, tapes and gasketing options.

Because of its location in the building enve-lope, the membrane air barrier cannot be accessed after construction. Therefore, spe-cific steps must be taken to ensure its long term effectiveness. As illustrated in Figure 4.6, standard installation practices include:

- Ensure that polyethylene is located on the warm side of the wall assembly at all times.
- Use UV stabilized, 0.15mm (6 mil) poly-ethylene (CGSB approved CAN2-51-34M86).
- Ensure compatibility of sealants with the polyethylene.
- Protect polyethylene from long-term exposure to sunlight.
- Minimize seams in the polyethylene by using the largest sheets possible.
- Overlap seams of polyethylene by a minimum of 150 mm (6 in.).
- Ensure seams are located over rigid backing (commonly framing materials).
- Caulk polyethylene seams with a flex-ible, non-drying sealant (such as acous-tical sealant) and staple seams through the sealant into solid backing. Avoid putting staples anywhere other than seams and edges.

- Sandwich polyethylene seams between two solid materials (e.g. stud/drywall, stud/strapping etc.).
- Ensure that polyethylene is not in direct contact with heat-producing surfaces (e.g. chimneys, hot water pipes, base-board heaters etc.).

4.5.2 Rigid Air Barrier Approach

The rigid air barrier approach was devel-oped as an alternative to the use of polyeth-ylene. The approach stresses the use of rigid materials (usually drywall) as com-ponents of the air barrier system, offering enhanced durability and structural rigid-ity.

Promoters of the approach point to re-duced installation costs and improved air barrier performance. By sealing the drywall at all junctions and by incorporating subfloors, rigid insulations and framing members as part of the system an effective air barrier can be achieved.

The rigid air barrier offers several advan-tages over its membrane counterpart. These include:

- It is not easily damaged in construction and can be easily repaired.
- A rigid air barrier will stand up to high wind pressures on the wall and ceiling.
- As the air seal is visible to the interior, a rigid air barrier can be repaired after construction.
- The vapour diffusion retarder can be provided by a variety of means (poly-ethylene, foil backing on drywall, or low permeance interior paints. Of note, while its use is largely redundant, many jurisdictions still require polyethylene as a VDR in the wall assembly).

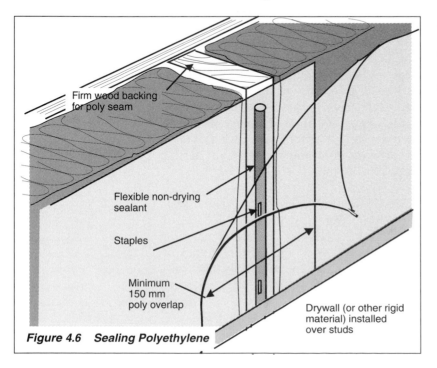

Firm wood backing for poly seam

Flexible non-drying sealant

Staples

Minimum 150 mm poly overlap

Drywall (or other rigid material) installed over studs

Figure 4.6 Sealing Polyethylene

Again, while many examples of rigid drywall details are provided in Part 2 of this book and Chapter 5 provides information on the materials used in this system, there are some general installation procedures to bear in mind:

- Seams between all components of the air barrier must be adequately sealed.
- Drywall is usually sealed to the framing using adhesive backed foam tape. The tape must be soft and resilient so that it will accommodate movement and fill all gaps. Most commonly used is a 1/2" by 3/16" low density closed cell PVC foam tape. This tape must be stapled to wood framing to ensure it will stay in place during drywalling. It is important that the drywall be screwed or nailed at an 8" spacing over the tape to provide an air tight seal.
- Caulking must be flexible, last the life of the building and adhere well to wood, drywall, foam insulations and metals.
- One part urethane caulkings have the best performance, but should not be used where chemical sensitivity is a concern. Flexible panel adhesives can successfully be used between framing members. As an alternative to one part urethane caulking, a plain formulation silicone caulking can be used but with a reduction in performance quality.

Figure 4.7 shows typical installation procedures for rigid air barriers.

Although a rigid air barrier minimizes the movement of indoor water vapour into insulated cavities by air leakage it does not act as a vapour diffusion retarder. It typically controls 70 to 90% of the moisture movement while the remaining 10 to 30% must be controlled through the use of a VDR.

The majority of rigid air barriers are installed with a vapour barrier primer or paint used as the VDR.

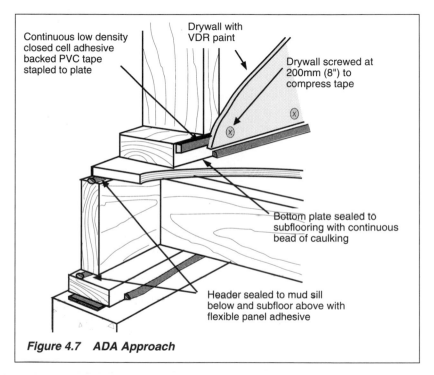

Figure 4.7 ADA Approach

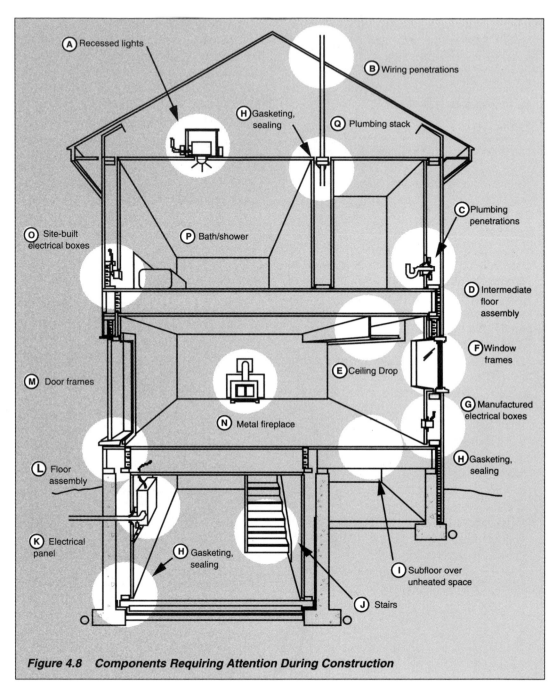

Figure 4.8   Components Requiring Attention During Construction

The greatest impediment to the widespread adoption of the rigid air barrier approach, are the changes in construction sequencing required.   Figure 4.8 highlights components of a typical house which require special attention at various stages of construction with regard to establishing a continuous rigid air barrier.

A Recessed lights
- Frame and seal airtight plywood box for recessed lights.
- Seal all wiring penetrations after light has been installed in box.
- Place foam gaskets on edge of box to act as seal to ceiling drywall.

B Wiring penetrations
- Seal wiring penetrations through partition end studs and partition top plates beneath insulated ceilings.

C Plumbing penetrations
- Frame airtight boxes for plumbing penetrations at exterior walls under sinks etc.
- Insulate rear of box with rigid insulation and caulk all joints. Place foam gaskets on face of box.
- Seal plumbing penetrations of box.

D Intermediate floor assembly
- Glue header to top plate below and subfloor above if it is to form air barrier.
- Place foam gasket and staple to upper edge of wall bottom plates, caulk bottom plates to subflooring.
- Place foam gasket and staple to exterior wall top plates.

E Ceiling drop
- Place oriented strand board (OSB), plywood or drywall behind future ceiling drop and gasket to top plate.

F Window frames
- Seal window frames to rough openings with expanding foam or with caulking and backer rod. Place foam tape on face of rough opening.

G Manufactured electrical boxes
- Install air tight electrical boxes designed for use with ADA.
- Caulk wiring penetration of electrical boxes.

H Gasketing and Sealing
- Place foam gaskets on upper edge of all exterior wall bottom plates and staple, seal bottom plates to basement slab or subflooring with caulking.
- Place foam gaskets on lower edge of all exterior wall top plates and staple.
- Place and staple foam gaskets on lower edge of both sides of partition wall top plate and partition end studs at exterior walls.

I Subfloor over unheated spaces
- Glue all subfloor joints over unheated spaces.
- Seal plumbing traps to subflooring when located over unheated spaces.

J Stairs
- Place oriented strand board (OSB), plywood or drywall behind stair stringers on exterior walls.

K Electrical panel
- Place oriented strand board, plywood or drywall behind future location of electrical panel on exterior wall and gasket to top plate.
- Install main electrical service panel on backer. All wiring surface is mounted and kept inside header joist air seal.
- Seal wiring penetrations of backer board.

L Floor assembly
- Place foam blocking between joists perpendicular to exterior walls. Caulk to top plate below, joists and subflooring above.

M Door frames
- Seal door frames to rough openings with expanding foam or caulking and backer rod. Place and staple foam gasket over inside face of rough opening.

N Metal fireplace
- Place drywall behind future location of metal fireplace. Gasket at bottom plate, seal at edges.
- Place metal flange for flue penetration.

O Site-built electrical boxes
- Frame airtight boxes for electrical outlet locations for conventional electrical boxes. Caulk all corners of framed air tight box and place and staple foam gasket over face of box.
- Place electrical boxes inside framed, insulated and sealed boxes.
- Seal all wiring penetrations of box.

P Bath/Shower
- Place oriented strand board (OSB), plywood or drywall behind future location of bath tub or shower. Gasket to bottom plate and seal at edges.

Q Plumbing stacks
- Place sheet rubber gaskets on plumbing stacks to seal to underside of partition wall top plates. Caulk and staple gasket to top plate.

Figure 4.8 Components Requiring Attention During Construction continued

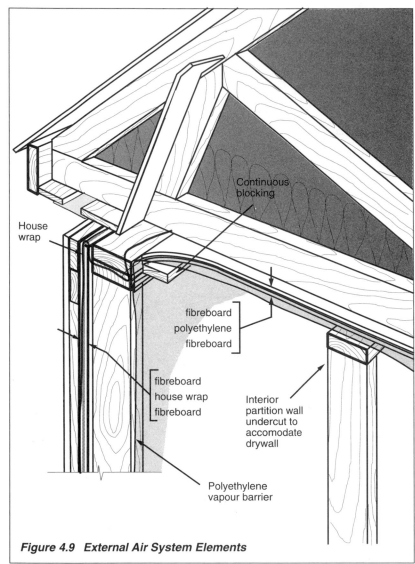

Figure 4.9 External Air System Elements

Labels in figure: House wrap; Continuous blocking; fibreboard / polyethylene / fibreboard; fibreboard / house wrap / fibreboard; Interior partition wall undercut to accomodate drywall; Polyethylene vapour barrier

4.5.3 Exterior Air Barrier System

Installation of the air barrier on the exterior of the framing assembly is a more recent application. An exterior air barrier can consist of certain rigid insulation materials, or more often a membrane house wrap such as spunbonded polyolefin or spunbonded polypropylene sheeting.

Because of their location on the exterior of the house these materials must possess the ability to allow adequate vapour transmission while stopping air movement, or be located so as to prevent condensation of moisture in the wall.

Exterior air barriers have the advantage of eliminating the labour required in sealing interior air barrier penetrations on walls and ceilings.

Some shiplap joint, rigid insulation materials have received approval as an accepted air barrier system. Where the wall cavity is maintained above the dew point, condensation problems are minimized.

If a membrane air barrier is used, seams in sheets are taped with a compatible material as specified by the manufacturer. The membrane air barrier can be sandwiched between rigid materials to provide protection against wind pressures. Typically fibreboard is used for this purpose. The air barrier must be made continuous and must be properly sealed to both the sill plate and the roofing members. Figure 4.9 illustrates one exterior air barrier system.

4.5.4 Combined Approach

Because both the membrane and rigid drywall approaches are combinations of many of the same components it is possible to combine them. For example, a hybrid air barrier may consist of drywall sealed to an exterior air barrier (header wrap) material which is carried around the floor assembly to join to the rigid air barrier below or above.

Chapter 5

MATERIALS

In recent years, there have been a number of developments in the materials and methods used to control heat loss and moisture movement in houses. This chapter discusses the various insulation, sealant and barrier materials available and their important characteristics. Later chapters discuss how these materials can be used to build quality houses.

This chapter is intended to provide general information only, to help you make planning and design decisions. For detailed information on particular materials, you should consult the technical literature or publications available from product manufacturers and distributors.

5.1 INSULATION MATERIALS

An insulating material slows the rate of heat flow from a warmer to a cooler area. Building envelopes generally comprise several components that act in different ways to slow heat flow.

- Most insulation materials consist of a cellular structure of a solid material that blocks heat flow by *radiation*, i.e. it is "opaque" to radiation much as dark glass is opaque to light transfer.

- The material contains tiny pockets of air or other gas(es) that reduce the *conduction* of heat.

- The air or gas pockets should be small enough that the possibility of heat flow via *convection* is reduced.

The thermal resistance of insulation materials will vary. It depends on:

- *cell structure* — the smaller the cells, the more effective they are in reducing heat transfer
- *the gas contained in the insulation* — some gases, such as refrigerant gases, have proven to be more effective than air at stopping heat transfer by conduction, and
- *moisture content* — any water trapped in an insulating material will tend to fill the spaces that are normally occupied by air or gas, reducing the material's ability to block heat flow by conduction.

However, heat transfer through the building envelope is probably most affected by the way the insulation is used and installed. The insulated space must be completely filled with material: there must be *no gaps or voids*. The material must be kept *dry*.

Loose fill insulations in particular must be installed according to design specifications if they are to perform as expected.

Insulation materials available include batt-type, loose fill, boardstock, spray-type, and radiant barriers. Figure 5.1 summarizes the characteristics, advantages and disadvantages of these various types of insulation.

5.1.1 Batt-type

Batt-type insulation is made from *glass* or *mineral fibres*. These fibrous materials are suitable for interior use, specifically inside exterior walls and foundation walls.

Insulation values of the various products are a function of the density of the materials. Increased density will generally improve the resistance of the materials and will at the same time reduce convective air movements within a building cavity.

The performance of batt-type insulation products is directly related to installation practices. Gaps around wiring and plumbing must be prevented, and batt materials must fill cavities completely and evenly.

5.1.2 Loose Fill

The *glass fibre* and *mineral wool* insulations are also available as loose fill.

Loose fill insulation is generally chopped, mixed with air, and blown into place using special machinery. It can be used in attics and inside above-grade exterior walls. It must be applied at the correct density to get good performance: consult manufacturers' literature for specific information on the recommended application procedure and density for a given product.

Cellulose fibre insulation is made from recycled newsprint. The raw material is shredded and treated with chemicals to control flammability, to prevent the growth of moulds and fungi, and to keep rodents from nesting in the material.

The chemicals can be added to the paper either dry or in a fine spray. (Note: as with any chemically treated substance, the builder should be aware of any indoor air quality problems that may result from the use of cellulose fibre — especially when building for people with heightened chemical sensitivities. In this case, cellulose fibre insulation used on the outside of the air barrier should not pose any problems.)

5.1.3 Boardstock

There are five types of rigid and semi-rigid boardstock insulations.

- *Expanded polystyrene* is made by expanding polystyrene beads in a mould. To make boardstock, large blocks of expanded polystyrene are cut into sheets of various thicknesses using hot wires.

  Low-density polystyrene (Type I) is suitable for use as a sheathing material for above-grade exterior walls. It can also be used to insulate interior basement walls, flat roofs and cathedral ceilings. High-density (Type II) material will withstand fairly high pressures, so it can be used to insulate below-grade walls on the exterior.

  Both materials can be used to insulate interior walls above and below grade level. Because these materials are combustible, they must be covered with a fire-protective covering such as 12.7 mm (1/2 in.) drywall if used to insulate living spaces.

- *Extruded polystyrene* is manufactured by extruding a hot mass of polystyrene through a slit. At atmospheric pressure it expands, creating a closed-cell foam material. This product is available as Type II, III and IV insulation and is suitable for use in all of the applications listed above for Type II expanded polystyrene.

Type IV is also suitable for use in built-up roofing applications (the higher the type number, the higher the density of the material.). As all cellular plastics are combustible, this material must be covered with a fire-protective covering if used in living spaces.

- *Rigid glass fibre* insulation can be used in built-up roofing applications and as exterior wall sheathing. It can also be used as below-grade exterior wall insulation: in this product, the fibres are aligned vertically so that any water that penetrates the insulation will run down the fibres; the density of such a product is three to five times more than a batt-type product.

- *Polyurethane and polyisocyanurate* are chemical reaction products of poly-alcohols and isocyanates. These insulations are made in continuous slabs which are then cut with hot wires. Polyurethane and polyisocyanurate boardstock can be used in all of the applications listed above for Type II expanded polystyrene. As noted earlier, these cellular plastics are combustible and must be covered with a fire-protective coating such as 12.7 mm (1/2 in.) drywall if used to insulate living spaces.

- *Phenolic insulation* is manufactured in much the same way as polyurethane and polyisocyanurate insulation. It is, however, much less combustible. Phenolic insulation is suitable for use as wall sheathing and for use inside, both above and below grade.

5.1.4 Spray-type

Spray-type insulations are a relatively recent innovation in the residential construction industry. There are currently three different types available.

- *Spray cellulose insulation* is available in a variety of formulations to suit specific applications. The material is applied using special applicators that mix the insulation material with adhesives, allowing it to hold together and adhere to the surface to which it is applied. Wet spray materials are gaining broader market acceptance because they offer thorough cavity coverage, reducing envelope air leakage characteristics. Several of the spray-applied materials require the installation of a mesh material over the face of the wall to contain the insulation material prior to the installation of the finished wallboard.

- *Two-component isocyanurate foam* is best suited for use in exterior stud wall cavities, in perimeter joist spaces, and in the shim spaces around windows and doors. Special applicators are used which mix the chemicals in the correct proportions.

- *Polyurethane* formulations are available for use in a variety of spray applications. For large applications, the material is mixed on site using special foaming equipment. For the smaller applications, single-component polyurethane foam is available in cans with "gun type dispensers" or in 4.5 kg (10 lb.) canisters for sealing shim spaces around windows and doors.

5.1.5 Radiant Barriers

A *radiant barrier* is a sheet of reflective material that is installed between a heat-radiating surface such as a roof heated by the sun and a heat-absorbing surface such as a ceiling. The reflective material stops the radiant transfer of heat between the two surfaces. Radiant barriers can be used to reduce cooling loads in summer by reducing the radiation of heat from the attic through the ceiling. The RSI (R) value provided by a radiant barrier depends on the direction of the heat flow.

Builders should take careful note before considering installing a radiant barrier. The "jury is still out" on whether the benefits of the system outweigh its limitations.

Insulation Material	RSI/mm (R/in.)	Density kg/m³ (lb/ft³)	Permeance ng/Pa·m²·s (grain/ft²·hr (in. Hg))	Flame Spread	Smoke Development	Advantages	Disadvantages
Batt-type:						• widely available • relatively noncombustible	• moisture and air infiltration reduce performance • compression due to weight of multiple layers of batts can reduce performance
Glass Fibre	0.022 (3.2)	24-40 (1.5-2.5)	1666 (29)	15	0	• see above	• see above
Mineral Fibre	0.024 (3.5)	24-64 (1.5-4.0)	1666 (29)	15	0	• see above	• see above
Loose Fills:						• widely available • relatively noncombustible	• settling and voids can occur if not applied at the correct density
Glass Fibre	0.020 (2.9)	9.8-40 (.061-2.5)	1666 (29)	15	0	• see above	• moisture and air infiltration reduce performance
Mineral Fibre	0.023 (3.3)	24-64 (1.5-4.0)	1666 (29)	15	0	• see above	• see above
Cellulose Fibre	0.025 (3.6)	35-48 (2.2-3.0)	1666 (29)	75	15	• more impervious to air flow if installed at high density (56 kg/m³ 3.5 lb/ft³)	• settles if exposed to heavy moisture for prolonged periods
Boardstock:						• variety of facing materials can provide air and moisture resistance and protection against ultraviolet rays	
Expanded Polystyrene (Type I and II)	0.026 (3.8) to 0.030 (4.4)	14.4-25.6 (0.9-1.6)	115-333 (2.0-5.8)	200	500	• available in several densities for use in different applications • moisture resistant • resistant to air flow	• combustible

Figure 5.1   Insulation Materials: Summary Information

Insulation Material	RSI/mm (R/in.)	Density kg/m³ (lb/ft³)	Permeance ng/Pam²s (grain/ft²hr (in. Hg))	Flame Spread	Smoke Development	Advantages	Disadvantages
Boardstock (con't):							
Extruded Polystyrene (Type III and IV)	0.034 (5.0)	25.6-32 (1.6-3.4)	23-92 (0.4-1.6)	200	500	• moisture resistant • resistant to air flow	• combustible
High Density Glass Fibre	0.029 (4.2) to 0.031 (4.5)	64-144 (4.0-9.0)	1725 (30)			• relatively noncombustible • can provide drainage next to the foundation	
Polyurethane	0.041 (6.0)	25.6-32 (1.6-2.0)	69 (1.2)	200	7500	• moisture resistant • resistant to air flow	• combustible
Phenolic	0.034 (5.0)	40 (2.5)		<25	<25	• noncombustible	• not readily available • not moisture resistant
Spray-Type:							
Cellulose Fibre	0.024 (3.5)	varies		<25	<25	• resistant to air flow	• must be installed by specially-trained contractors
Isocyanurate	0.034 (5.0)	13.6 (0.85)	575 (10)	<25	<300	• see above	• see above
Polyurethane	0.041 (6.0)	varies		<25	up to 500	• see above	• see above • combustible

Figure 5.1   Insulation Materials: Summary Information

5.2 VAPOUR DIFFUSION RETARDERS (VDRS)

5.2.1 Polyethylene Films

Polyethylene film is very resistant to the flow of water vapour (see Appendix II), and consequently is suitable for use as a vapour diffusion retarder. As discussed in Section 5.3, it can also be used as an air barrier.

VDR polyethylene should comply with the Canadian General Standards Board (CGSB) standard for upgraded 0.15 mm (6 mil) polyethylene (CAN2-51-34M86) — check the label to be sure. If the film is made from virgin polyethylene resins, it will be quite clear. If it is cloudy, it probably contains reused resins. There have been some concerns that polyethylene made from reused resins may break down when exposed to ultraviolet radiation (sunlight). Studies to date have not found this to be much of a problem, but it may be best to use polyethylene made from virgin resins just to be on the safe side.

Choose cross-laminated film for best performance. Whatever product you choose must be installed carefully to ensure reliable long-term performance.

5.2.2 Paints

Paint can also be used as a vapour diffusion retarder. As shown in Appendix II, varying numbers of coats of different types of paints will provide the required resistance to the flow of water vapour. Using paint as the vapour diffusion retarder can have advantages because it is applied at the end of the construction process and serves as the finish as well. However, care is required to ensure adequate coverage to attain a suitable perm rating. If you're planning to use paint as the VDR, ensure that it meets the requirements of the CGSB standard *Method for Permeance of Coated Wallboard* (CAN/CGSB-1.501-M88).

5.3 AIR BARRIERS

5.3.1 Air Barrier Materials

As noted in Chapter 4, air barriers can be either membrane or rigid barriers. Polyethylene film is the most commonly used membrane air barrier material. In most cases, it also functions as the VDR. It must be properly situated in the wall or ceiling assembly and sealed with compatible materials (see details in chapters 8 and 9). In the rigid air barrier approach, drywall, wood panel sheathing or certain boardstock insulations such as extruded polystyrene, foil-faced polyurethane and polyisocyanurate boardstocks can be used as the air barrier, as long as they are sealed with appropriate gaskets and sealants (see details in chapters 8 and 9). Where low permeability materials are installed on the exterior of the building, a Type I air barrier (polyethylene) must be installed on the building interior.

5.3.2 Gasket Materials

Gaskets can be used to seal joints between materials to prevent air leakage. They are most commonly used in the sill plate area but are also used extensively to seal rigid air barriers. The method of attachment must be compatible with the gasket profile and must not compromise the integrity of the air seal.

Desirable characteristics of gaskets are outlined below.

* Gaskets must be sufficiently resilient to maintain their seal while accommodating changes in the dimensions of the materials they are intended to seal. Changes in dimensions can result from seasonal changes in building frame moisture content, building movement due to settlement, and wind and snow loads.

- They must be thick and wide enough to fill and effectively seal the cavity between the two materials and must be sufficiently compressible to not cause visible deformations in interior cladding materials

- Gaskets must be made of a material that does not degrade or break down over the service life of the building. They should be made of material that can be installed under extreme weather conditions.

- Gaskets should not be water soluble nor give off toxic fumes. They should not chemically attack or be affected by the building materials they join or abut.

- The gaskets must be sufficiently durable to withstand the construction process and have a profile that does not lose sealing capacity over time. They must be able to withstand the wind loads and other pressure forces that are expected to act on them.

Materials which have been employed effectively as gasketing include open and closed cell polyethylene, butyl glazing tapes, neoprene, and polymer type gaskets.

5.3.3 Caulking and Sealant Materials

Caulking is used to seal joints between materials to prevent air leakage and penetration by snow and rain. Most caulks should be applied at temperatures above 15°C (59°F). Manufacturers provide directions for application and storage.

A wide variety of caulking materials can be used with varying degrees of success in a given application. The following list provides some guidelines. (Note: builders need to be sensitive to possible indoor air quality problems that can result from the use of a particular sealant—especially in housing for the chemically sensitive.)

- *Oil- or resin-based* caulking will bond to most surfaces. However, it is not paintable or durable as it shrinks and hardens.

- *Latex-based* caulking bonds to most surfaces. It does shrink, but is more durable than oil- or resin-based caulking.

- *Butyl rubber* will bond to most surfaces and is paintable. It is particularly suitable for use on masonry, but not where the joint expands and contracts. It is paintable, but shrinks and has limited durability.

- *Nitrile rubber* bonds well to masonry and metal, but not to painted materials. It is paintable and suited for use in areas where moisture is a problem. It shrinks appreciably but is considered to be durable.

- *Neoprene rubber* will bond to most surfaces, and is particularly suitable for use on concrete. It is paintable and considered durable.

- *Silicone* caulking bonds to most surfaces except concrete. It is paintable, does not shrink, and is considered to be very durable. It is also suitable for use in areas where moisture is a problem.

- *Polysulfide* caulking bonds well to primed surfaces. It doesn't shrink and is considered to be very durable. It should not be left exposed.

- *Polyurethane foam* sealant will bond to most surfaces. It does not shrink and is considered to be very durable.

- *Hypalon* caulking must be applied after a primer if the substrate is porous. However, it does bond to most surfaces. It is paintable, does not shrink, and is considered to be very durable.

- *Acoustical* sealant does not require a primer and bonds well to polyethylene. It is not paintable and cannot be used in areas where it will be left exposed because it does not harden. It is considered to be very durable. (To date, this has been the sealant of choice for polyethylene.)

5.3.4   Weatherstripping Materials

This section provides information on five types of materials used in the manufacture of weatherstripping — the "active" material in the product. Weatherstripping also usually includes a fastening system. Fastening systems are made from wood, vinyl, or metal, or may simply be grooves cut into the window or door frame.

The effectiveness of the seal depends on the weatherstripping material used, the fastening system and the design of the seal. Hinged units, such as casement windows and entrance doors, work on the principle of compression to create a seal when closed. Sliding units, such as sliding patio doors and horizontally sliding windows, use a friction seal. In general, compression units provide a better seal than friction units as long as there is no warping.

Before purchasing and installing a window or door system, you should assess the effectiveness and durability of the seals and designs used. Ask for information on air leakage characteristics and long-term performance before selecting a system.

- *Neoprene rubber* is used in compression seals. This material is often used in automobile door gaskets. It has good "memory" and compression set resistance: the ability to return to its original shape. It offers superior resistance to the effects of heat, cold, oxidation and ultraviolet light. It does not shrink or crack. It also has excellent impact resistance and good abrasion resistance.

- *PVC, TPR and polypropylene plastics* do not perform as well as rubber products. PVC may shrink or lose flexibility over time. It can be affected by heat, cold and ultraviolet light. If exposed to high temperatures, which sometimes occur between glazing panels, the PVC can set in the compressed form and not return to its original shape.

Polypropylene has the best performance characteristics of the three. These materials are generally used in compression applications; some plastics are also used in friction seals.

- *Steel, bronze and brass* do not shrink and are not affected by environmental conditions. They are generally used in friction applications and are very durable. However, since they are easily deformed and rendered ineffective, the design of any window or door system using metal strips is critical.

- *Pile* is particularly suited for use in sliding seals. It is very effective when dry and new, but wears and loses its memory. Some pile products incorporate a central fin which improves durability and performance.

- *Magnetic strips* are two-part systems. A metal strip is attached to the frame of the window or door. A magnetic strip held in vinyl is attached to the face of the window or door. When the unit is closed, the metal and magnet come together to form the seal. This system works well in moderate climates, but the fastening strips perform poorly in cold weather.

5.3.5   Specialty Items

Single component polymeric foam is especially good for sealing sill plates, rough-frame openings around windows and doors, and other openings too large to be sealed with caulking or gasket materials.

5.4 WEATHER BARRIERS

As noted in Chapter 4, weather barriers are materials placed on the exterior of the wall or ceiling assembly to protect the system from the elements. They resist the harmful effects of rain and snow, keeping the insulation and structural members dry. They also improve the thermal performance of the system by keeping wind out.

5.4.1 Spunbonded Polyolefin and Woven Polypropylene

These two materials are relative newcomers on the housing scene. They shed liquid water and resist air flow but are permeable to the diffusion of water vapour. They are tough enough to withstand severe wind and the rough handling they might receive during the construction process. However, they should be covered within 60 days to prevent potential deterioration caused by ultraviolet radiation. Special sheathing tape is used to seal seams, holes, and openings around windows, doors and service penetrations.

5.4.2 Perforated Polyethylene Films

When water vapour enters the building envelope from inside the house, it can get trapped there by continuous polyethylene films. Perforated materials were developed to overcome this problem. Seams, holes and openings around windows, doors and service penetrations can be sealed with patches of the polyethylene material or a sealant compatible with it.

5.4.3 Building Papers

Asphalt-impregnated building papers have been used for decades. They must be lapped and secured to perform adequately. Strapping is most commonly used to hold the paper in place while the siding is applied. Both solid and perforated building papers are available.

5.5 MOISTURE BARRIERS

A moisture barrier was defined in Chapter 4 as a material, membrane, or coating used to keep moisture which might come through the foundation wall from penetrating the interior foundation insulation or the wood members supporting the insulation and interior finishes. It is installed only on the below-grade portion of the foundation walls or under floor slabs.

Materials that can be used as moisture barriers include:

- asphalt emulsions applied to the interior surface of the wall to provide a continuous membrane — tar is not recommended because it often produces an objectionable odour.
- sheet materials, such as 0.05 mm (2 mil) polyethylene or building papers. All seams must be overlapped and sealed. Sheet materials can be fastened to the wall using adhesives or a cold asphalt mastic such as plastic roof cement. Alternatively, they can be held in place by furring strips secured to the wall.
- specially designed foundation membranes that both provide moisture protection and free flow drainage capabilities.

5.6 DAMPPROOFING MATERIALS

To prevent water problems with foundations:

- ensure that the footing and sub-slab area is properly drained
- use granular fill to backfill the wall
- grade the earth around the house away from the building
- use eavestroughing and downspouts to take rainwater away from the area next to the foundation, and
- dampproof the wall from grade level to the bottom and over the joint with the footing.

The material most commonly used for this dampproofing is a spray- or mop-applied bituminous substance. It fills the pores in the concrete, reducing the penetration of liquid water due to pressure against the wall and capillary action. This material has been used by builders for many years. It is easy to apply, effective, and very durable.

5.7 SUMMARY

As this chapter has indicated, there are a number of suitable barrier and insulation materials on the market. The following chapters look at how these various materials can be incorporated into the building envelope to produce durable and comfortable houses.

Part 2

BUILDING ENVELOPE

Chapter 6

FOUNDATIONS

This chapter discusses how to reduce heat losses from foundations and to minimize air flow and moisture problems in this part of the building. It includes specific techniques for different types of foundation systems.

In deciding what type of foundation to use, you should consider the following factors:

- local water table and soil conditions; for example, expanding or regular clay, sand, gravel, rock, and presence of contaminants
- availability of materials
- cost and ease of construction
- local building codes
- market acceptance of basement vs. slab-on-grade foundations, and
- market acceptance of materials; for example, poured concrete vs. concrete block vs. preserved wood.

Once you have made your decision, use the techniques outlined in Sections 6.4 through 6.8 to build a durable foundation with minimal heat loss.

6.1 HEAT LOSS THROUGH FOUNDATIONS

Heat loss from the foundation can account for a significant amount of the total heat loss in the average house. The rate of heat loss will be directly proportionate to the air and ground soil temperature, and to the amount of moisture in the surrounding soils. Figure 6.1 shows the various ways in which this heat loss may occur.

Most losses occur because of:

- heat loss by air leakage through the sill plate/header/subfloor assembly
- heat conduction through above-grade sections of foundation walls, and
- heat conduction below grade to the surrounding soil and to ground water.

These mechanisms of heat loss are discussed in more detail below.

6.1.1 Heat Loss Above Grade

Heat loss through foundation walls above grade is directly affected by the temperature difference between inside and outside, and the resistance to heat loss of the foundation wall. Heat loss through the above-grade portions of foundation walls tends to be significant even though only a small amount of area is exposed to the outside air. This is primarily because the materials used for foundation walls have a lower insulating value than those used for the main building envelope.

6.1.2 Heat Loss Below Grade

Several factors influence heat loss below grade.

Depth
- Soil temperature increases with depth. Those parts of the foundation close to the surface have higher heat loss rates than those at the footings.

Presence of water
- A high water table within 600 mm (24 in.) of the slab will increase heat loss from the slab because water passing by the slab will carry heat away with it. Similarly, surface water draining to the weeping tile will carry away heat from the foundation wall.

6.1.3 Air Leakage

Air leakage can occur both above and below grade.

- Cold air will tend to enter through the lower portions of a house because of negative pressures created by stack action and combustion heating equipment (see Section 2.2).

- The foundation wall/sill plate/joist header junction has multiple joints and connections through which air can move.

- Cracks and joints in poured concrete and concrete block walls, and openings around service penetrations and windows increase the number of paths for air leakage.

- Air will leak in through the floor drain if it does not have a trap, or if the trap is dry. Air will also leak through the weeping tiles if they are connected to the sump hole.

Where the house is located in an area with a significant source of radon in the ground any below-grade crack or hole that allows air into the basement or crawlspace will also allow radon gas to enter. This air leakage may require preventative measures (see Section 2.6).

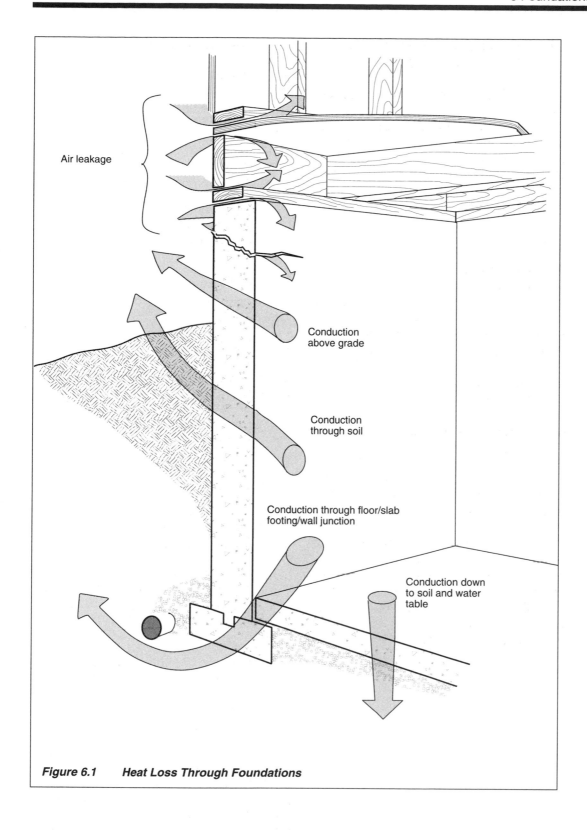

Air leakage

Conduction
above grade

Conduction
through soil

Conduction through floor/slab
footing/wall junction

Conduction down
to soil and water
table

Figure 6.1   *Heat Loss Through Foundations*

6.2 REDUCING HEAT LOSS

Proper insulation can greatly reduce heat loss from the foundation. Proper foundation insulation is not necessarily expensive, and it can improve thermal comfort, reduce surface condensation on walls, and add protection for the structure, as well as reduce energy costs.

The exact amount of insulation required and its location in each foundation design depends on local soil conditions, the method of construction, and construction costs. The following points apply to non-permafrost regions and should be applied to any given construction practice. (Specific techniques for constructing in permafrost regions are described in Section 6.8.)

• In energy efficient housing, full depth foundation insulation is now commonly accepted throughout the country.

• The basement walls, especially those near to and above grade, should be insulated to the same level as other exterior walls.

• It may also be possible to install extra insulation at the header area (see Figure 6.2).

• Unless the soil is wet or there is a high winter water table, it may not be economical to insulate basement slabs. However, insulation may be desirable for comfort reasons if the basement is to be a finished space. Floor slabs of heated crawlspaces should be insulated to RSI 1.8 (R-10) with a minimum 1.0 m (3 ft.) width of insulation around the entire slab perimeter.

• The foundation wall/sill plate/joist header junction should be air tight. Air tight construction details should be used throughout the rest of the foundation wall, floor, and wall/floor junction.

• Thermal bridging from a heated foundation wall to unheated sections of masonry veneer should be eliminated.

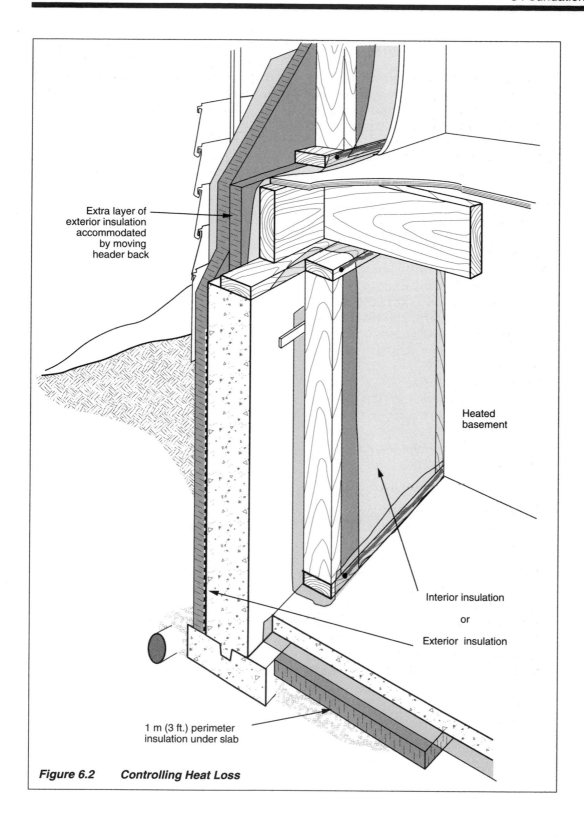

Extra layer of
exterior insulation
accommodated
by moving
header back

Heated
basement

Interior insulation

or

Exterior insulation

1 m (3 ft.) perimeter
insulation under slab

Figure 6.2   Controlling Heat Loss

6.3 REDUCING MOISTURE

Moisture forming on basement walls and floors and water leaking through the foundation from the outside can be major problems. Water can enter basements and crawlspaces in several ways.

- The curing of concrete in walls and slabs will release water, increasing indoor humidity levels.
- Water vapour may diffuse inward through improperly dampproofed walls and slabs.
- In summer, infiltrating air may carry in water vapour. Effective air sealing techniques should reduce this problem.
- Water can leak through cracks in the foundation. Proper construction techniques and exterior drainage can prevent leakage.

Insulation, dampproofing and drainage systems can reduce foundation moisture problems (see Figure 6.3).

6.3.1 Curing Concrete

As concrete cures, it releases large amounts of water vapour, especially in the first few months. If the interior surfaces of the foundation walls and floor are left uncovered, this moisture will be released into the house. If the concrete surfaces are covered, the moisture may move into other areas in the walls or floor, causing problems such as staining.

To minimize these problems, one should:
- allow concrete foundations to dry for as long as possible before placing the interior insulation and vapour diffusion retarder.
- place a moisture barrier over the inside surface of the concrete walls from the footing up to grade level.
- leave above-grade sections of the concrete wall uncovered as long as possible to allow moisture to escape.

6.3.2 Reducing Condensation

To reduce condensation on the interior of basement or crawlspace walls, follow these steps:

A) Insulate the concrete on the outside. This will keep it warm and reduce the likelihood of condensation.
B) Even though the foundation wall acts as an air barrier, you should place a continuous air barrier and a vapour diffusion retarder over the inside of all insulation when insulating on the interior.
C) Seal the bottom plate/floor slab junction to keep warm air from leaking behind the wall and condensing on the cold concrete.

6.3.3 Preventing Outside Water Penetration

To keep outside water from penetrating the foundation, follow standard good building practices. Use proper flashing around windows and doors to keep rain water away, and provide drain holes in brick veneer.

To safely remove ground water:
- slope the grade away from the house
- use free draining backfill and/or self draining exterior insulation
- use exterior dampproofing, and
- install eavestrough downspouts to direct water away from foundations.

As an alternative to free draining backfill materials, several membrane materials are gaining wider market acceptance. These materials provide a free passage for water to travel to the weeping tile, reducing pressures which force water through the foundation wall. A variety of products with different configurations are now available on the market. Self draining insulation materials are also available.

The remainder of this chapter describes detailed techniques for constructing durable foundations with minimal heat loss and moisture penetration.

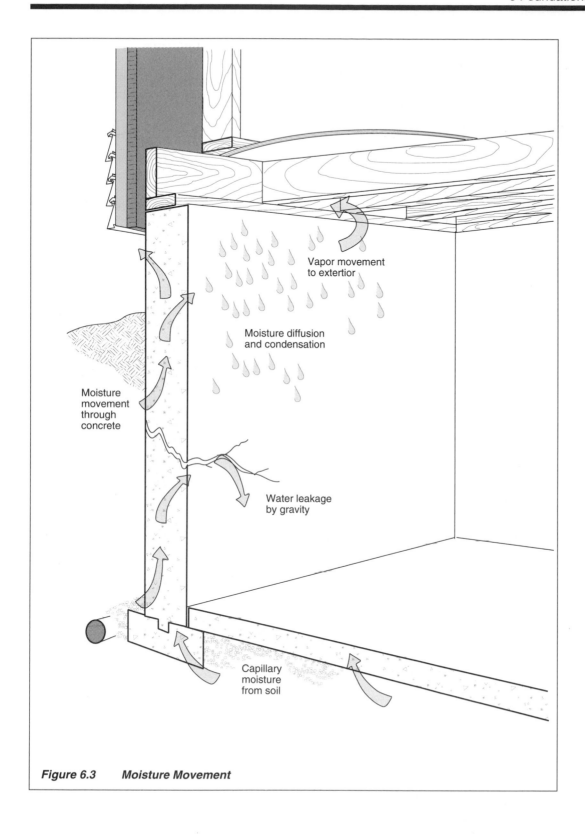

Vapor movement
to extertior

Moisture diffusion
and condensation

Moisture
movement
through
concrete

Water leakage
by gravity

Capillary
moisture
from soil

Figure 6.3   Moisture Movement

6.4 POURED CONCRETE, BLOCK BASEMENTS AND HEATED CRAWLSPACES

Basement and crawlspace foundations may be insulated either on the exterior or on the interior. Both approaches have advantages and disadvantages.

Exterior Insulation Advantages

- The wall is kept warmer, reducing condensation problems.
- There is reduced potential for adfreezing (where the soil freezes to the surface of the foundation and may lift it).
- There are reduced labour requirements.
- This approach allows for improved airtightness and reduced thermal bridging at the header joist.
- The thermal mass of the foundation wall is inside the house.

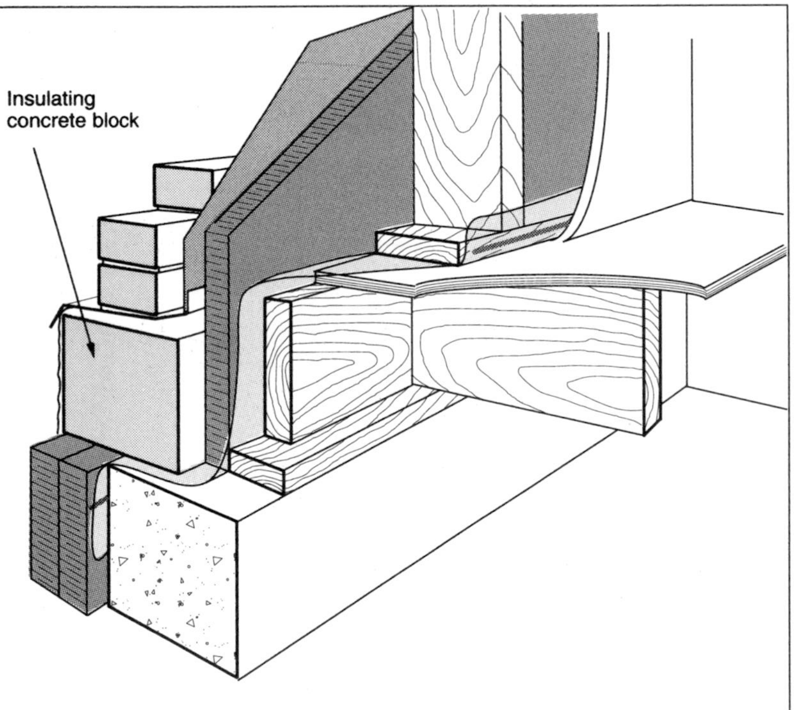

Exterior Insulation Disadvantages

- There may be higher material costs, including allowing for above-grade protection.
- This approach does not provide a finished interior surface.
- There is potential for thermal bridging from the warm foundation to the cold brick veneer.
- Extra care is required during backfilling.

Interior Insulation Advantages

- Insulation is provided as part of the interior finishing.
- There are no changes to exterior finishing details.
- Space is provided for mechanicals and electricals.
- Thermal bridging with brick veneer cladding is prevented.

Interior Insulation Disadvantages

- There is a potential for concealed condensation — this approach requires a continuous air barrier and a vapour diffusion retarder.
- The foundation wall is exposed to temperature fluctuations and there is potential for adfreezing.
- Sealing the sill plate/header joist assembly can be difficult.

You should consider all these factors when deciding whether to use exterior or interior insulation methods. If you are using brick cladding, you should definitely consider using interior insulation to prevent thermal bridging. Three alternative sill details for brick veneer are illustrated in Figures 6.4 to 6.6.

The detailed techniques required for the exterior and interior insulation approaches are summarized in the remainder of this section.

Insulating concrete block

Figure 6.4   Sill Detail: Brick Veneer on Concrete Foundation with Exterior Insulation

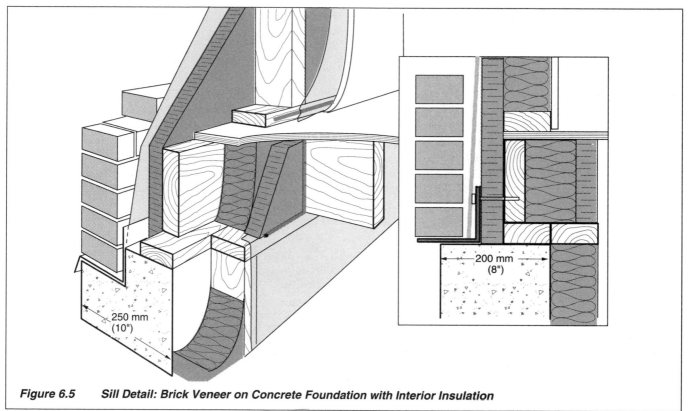

Figure 6.5 **Sill Detail: Brick Veneer on Concrete Foundation with Interior Insulation**

200 mm
(8")

250 mm
(10")

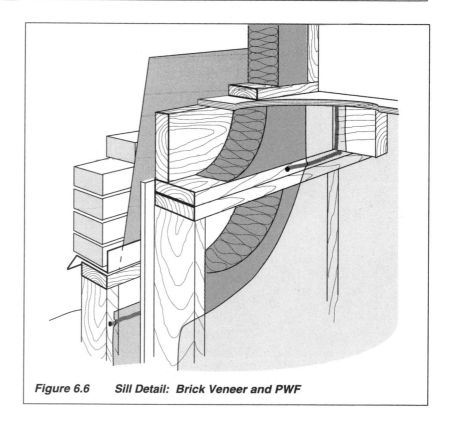

Figure 6.6 **Sill Detail: Brick Veneer and PWF**

6.4.1 Externally Insulated Masonry Foundations

Rigid insulating board can be affixed to the exterior basement wall using a variety of fastening techniques and protective coverings. It only needs to be attached above grade, because below grade the soil pressure will hold it in place. As with all foundation work, good dampproofing and drainage techniques are important for long-term performance. The amount of insulation installed will depend on regional climate, soil conditions, and desired energy performance (see Figure 6.7).

System considerations and special techniques for dealing with service penetrations through externally insulated masonry walls are summarized below.

System Considerations

Dampproofing

- 0.15 mm (6 mil) polyethylene dampproofing can be used as an air barrier by extending the material up the wall from the footing to the underside of the bottom wall plate.

- Some dampproofing materials will react with certain foam insulations while curing, so allow adequate drying time before installing board insulation.

Insulation

- Type II expanded polystyrene and Types III and IV extruded polystyrene are suitable for exterior insulation. Type II retains more moisture and should not be used in areas with poorly draining soils.

- Rigid glass fibre insulation can provide excellent drainage even without free-draining backfill if detailing at the footing allows water to flow into the perimeter drain or weeping tile.

- Both rigid polystyrene and glass fibre boardstock must be installed vertically.

Fastening

- Nailing strips can be cast into poured concrete foundations at locations dictated by the amount of exposure above grade. Use galvanized nails with plastic washers to hold the insulation in place.

- If above grade exposure is not significant, you can fasten insulation directly to the sill plate. It will then be held in place by soil pressure below grade.

- Several fastening systems are available for direct connection to masonry surfaces.

- Polystyrene can be placed in forms and cast in concrete.

Protective Covering

A variety of protective coverings are available. These include:

- galvanized wire lath and stucco — more fasteners may be needed to make the lath rigid
- fibre reinforced polymer modified cement or other foundation coating
- cement mill board, and
- pressure-treated plywood.

Use proper flashing techniques at the top of the insulation to keep water from leaking between the insulation and the foundation wall. Follow good backfill practices to ensure proper drainage.

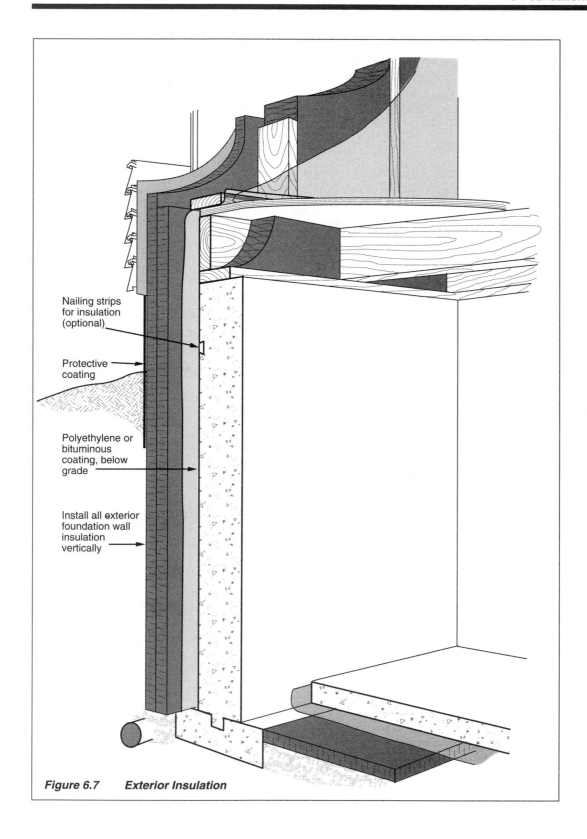

Nailing strips
for insulation
(optional)

Protective
coating

Polyethylene or
bituminous
coating, below
grade

Install all exterior
foundation wall
insulation
vertically

Figure 6.7   *Exterior Insulation*

Service Penetrations Through Externally Insulated Masonry Walls

Service penetrations must be carefully sealed to prevent air leakage.

Electrical Conduits

Electrical conduits should be sealed where they pass through the foundation wall or joist headers, and inside the conduit where they terminate in the panel. Be sure to follow all local codes and regulations.

- *Foundation wall penetration* (see Figure 6.8): seal with polysulphide or high solids content butyl.
- *Joist header* (see Figure 6.9): seal with an acrylic, silicone, or butyl sealant which is compatible with wood and with the conduit material.
- *Conduit sleeve end* (Figure 6.9): seal with a compatible caulking compound.

Air Supply and Exhaust Ducts

Sheet metal ducting required for dryers, heat recovery ventilator systems or make-up air may pass through the joist header. This ducting should be sealed with sheet metal and an appropriate sealant, or a rubber gasket (an elastomeric roofing membrane), or with acrylic, silicone, or high solids content butyl caulking material (see Figure 6.10). Consider the temperature of the duct when selecting the sealing material.

Plumbing

Seal around the below-grade water supply and drains in the same manner as electrical conduits. Copper piping should not be in direct contact with concrete because it may react with chemicals in the concrete.

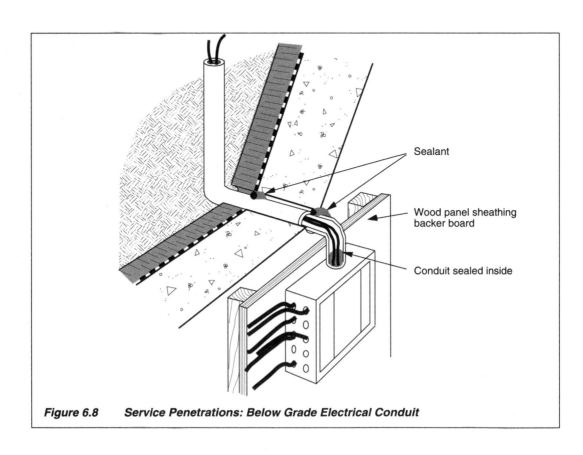

Figure 6.8   *Service Penetrations: Below Grade Electrical Conduit*

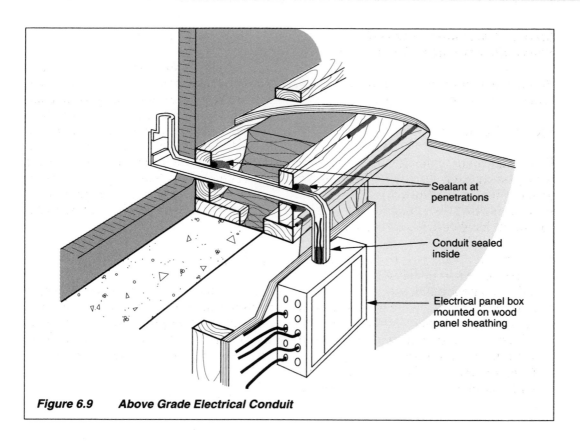

Figure 6.9   Above Grade Electrical Conduit

Sealant at penetrations

Conduit sealed inside

Electrical panel box mounted on wood panel sheathing

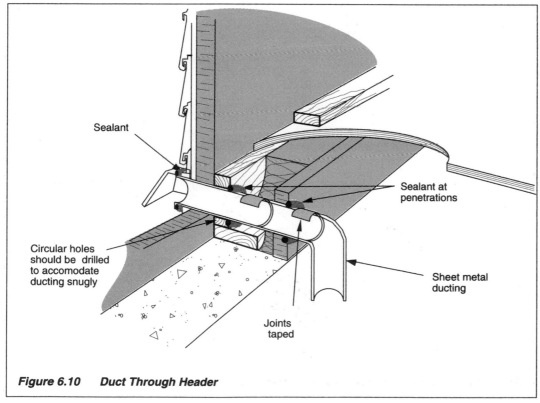

Figure 6.10   Duct Through Header

Sealant

Sealant at penetrations

Circular holes should be drilled to accomodate ducting snugly

Sheet metal ducting

Joints taped

6.4.2 Internally Insulated Foundation Walls

There are several different ways to insulate the foundation from the interior. Before installing any insulation, be sure to apply a moisture barrier to the inside of the foundation wall from grade level to the floor of the basement. This is in addition to, and does not replace, the normal exterior dampproofing. Where free draining backfill or materials are employed, clear access to weeping tiles must be maintained.

The thickness of insulation required will depend on the RSI (R) value of the particular product or system, the potential loss of interior space, ease of installation, regional climate, and desired energy performance (see Figure 6.11).

System considerations and special techniques for dealing with service penetrations through interior insulated concrete walls are summarized below.

System Considerations

Dampproofing

- To create the moisture barrier, apply polyethylene, building paper or a bituminous coating to the concrete foundation wall from grade to footing level.

Drainage

- Membrane drainage materials must be flashed to prevent above-grade water from entering behind the materials. Many of the systems require special detailing at the footing/foundation wall junction. Where free draining backfill materials are plentiful, they may prove less expensive than membrane drainage materials.

Insulation

- Any rigid board insulation — polystyrene Types I, II, III and IV — can be directly fastened to the foundation wall or placed between layers of strapping. Thick insulation will require either a metal channel or cross strapping to provide support for drywall.

- Frame walls can be constructed and spaced out from the foundation walls to accommodate thicker insulation. The wall can be nailed to joists above and fastened with concrete nails to the floor slab. It is often advisable to install batt insulation in two steps, with one layer horizontally between the frame and foundation and the other vertically in the stud cavity. This will reduce convection loops behind the studs where the batt insulation does not completely fill the cavity. Since the wall is not load-bearing, framing can be 38x64 mm (2x3 in.) lumber at 0.6 m (2 ft.) centres to keep costs down. RSI 3.5 (R-20) can be reached if the horizontal layer has a value of RSI 2.1 (R-12) and the layer between the studs is RSI 1.4 (R-8).

Air Sealing

- Polyethylene can be applied from the underside of the first floor joists to the floor slab as long as the sill plate/joist assembly is properly detailed. Seal the joints with either a gasketing material or a polysulphide sealant.

- Use sill plate gaskets to provide a seal between the sill plate and the concrete.

- One way to provide air barrier continuity at the floor joist assembly is to install a sheet of house wrap material over the top of the foundation wall gasket before the sill plate is installed (see Figure 6.11).

- Provide enough overlap with the foundation wall to wrap the sheet around the floor assembly and run it under the bottom plate of the first floor frame wall. This practice will make a continuous seal between the foundation wall and the above-grade exterior wall.

- It is also possible to use polyethylene in this location. Since the polyethylene will also act as a vapour diffusion retarder, the one-third/two-thirds rule must be followed. In northern regions, more insulation, up to four-fifths of the total insulating value, may be required on the exterior.

- Alternatively, place nonpermeable rigid board insulation or glass fibre insulation encased in polyethylene bags made for this purpose in each of the joist spaces and seal with caulking. These can then be connected to the polyethylene on the wall below using acoustical caulk and staples

- Spray-in-place foam insulations represent a viable alternative for efficiently sealing header assemblies.

- A rigid air barrier approach to air sealing an interior insulated foundation is illustrated in Figure 6.12.

Electrical and Plumbing

- To provide a space for electricals and plumbing in a finished basement, nail a layer of 38x38 mm (2x2 in.) strapping horizontally over the insulation system after polyethylene has been installed. Wiring and plumbing are then installed before the drywall is hung. Always locate plumbing on the warm side of the insulation.

- Where using drywall as the air barrier, ensure that a good air and vapour seal is provided to reduce the risk of moisture entering the insulated cavity.

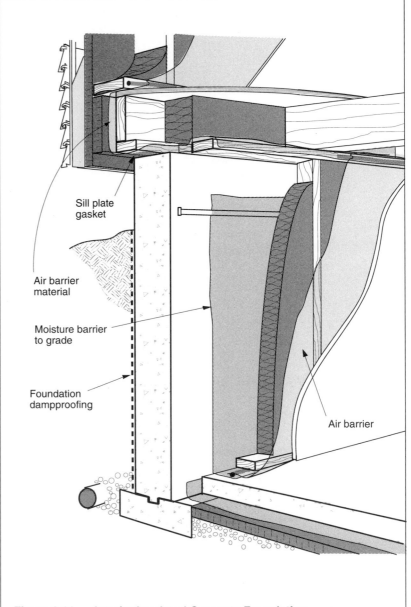

Sill plate gasket

Air barrier material

Moisture barrier to grade

Foundation dampproofing

Air barrier

Figure 6.11   Interior Insulated Concrete Foundation

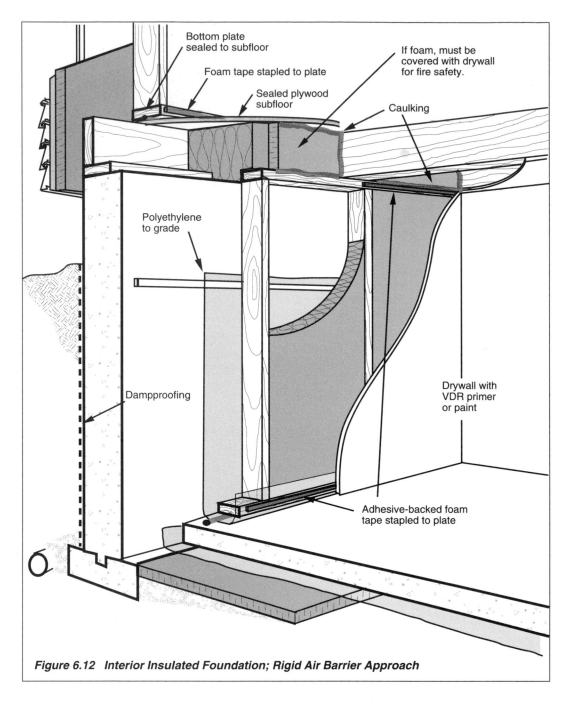

Figure 6.12 Interior Insulated Foundation; Rigid Air Barrier Approach

Service Penetrations Through Interior Insulated Concrete Walls

Although it may be necessary to provide service penetrations through concrete walls, conduits can often be placed in "softer" areas such as headers or suspended floor

assemblies as long as the design has taken this placement into account.

Be sure to locate service conduits, ducts and grilles where they won't be affected by drifting snow.

Electrical Conduits, Wiring, Gas Lines, and Plumbing Lines

These service penetrations must be sealed as they pass through the air barrier. There are several ways to do this.

- After the vapour diffusion retarder is in place, seal it to the plywood backing board with caulking and staples. Where the conduit or duct passes through the plywood, seal gaps with caulking.

- Mount main electrical service boxes directly on a plywood backing board and keep all wiring in front of the backing board, except those wires which run into the frame walls of the basement itself (see Figure 6.13). Where possible, try to keep these circuits in interior partition walls. Seal penetrations with caulking where wiring runs through the framing.

- Seal the entire header assembly with spray-in-place foam insulation.

- When using ADA, seal drywall to plywood with mud and tape.

Electrical Boxes

All electrical boxes must be sealed to the air barrier (see Chapter 8).

Windows

Seal the window frame to the air barrier or directly to the concrete foundation wall. Since most basement windows are mounted in the exterior foundation wall, the air barrier must span from the inside wall to the outside wall. This sealing can be done with a plywood window liner (see Figure 6.14). Alternative air sealing details are provided in Chapter 10.

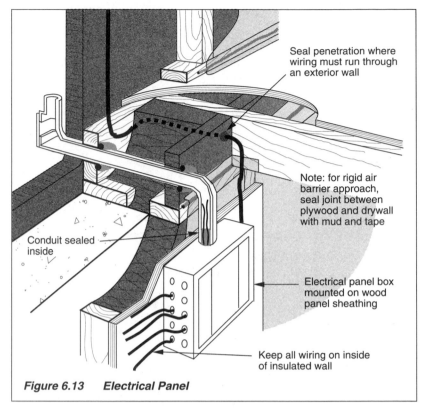

Figure 6.13   Electrical Panel

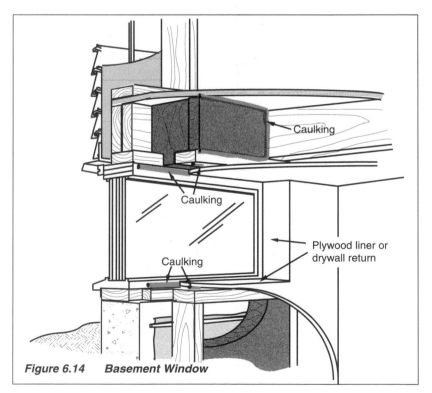

Figure 6.14   Basement Window

6.5 PRESERVED WOOD FOUNDATIONS

Preserved wood foundations (PWF) consist of a wood framed wall made of preservative pressure-treated lumber designed to be used below grade (see Figure 6.15). These foundations have performed successfully for more than 30 years, but are still not widely accepted in all areas of the country. If you are unfamiliar with PWF techniques, you should obtain background references. The Canadian Standards Association (CSA) publication, *Construction of Preserved Wood Foundations* (CAN3-S406), provides complete information on the permanent wood foundations.

Preserved wood foundations can be built using a variety of floor systems, including a concrete slab, a wood sleeper floor, or a suspended floor. They have a number of advantages.

- Preserved wood foundations can be built by a standard framing crew.
- They can eliminate the need for concrete, permitting cold weather construction.
- These foundations are especially useful where concrete is difficult to obtain or place.
- High levels of batt insulation can be installed in the foundation walls.
- Air sealing techniques are the same as for above-grade walls.
- Crawlspace basement floors can have high floor insulation levels using batt insulation.
- Wall height is not restricted by the height of concrete forms.

However, there are also disadvantages to using preserved wood foundations.

- Market acceptance is low in some areas.
- Additional site supervision may be required.
- The foundation must be carefully designed to resist the pressure of backfill.

- These foundations require specialized materials. Particular grades and dimensions of pressure-treated lumber or stainless steel fasteners, for example, may require lead time to order.
- Alterations to first floor joist framing are required to account for horizontal loads.
- The excavation must be carefully designed if pressure-treated wood footings are used.
- A gravel drainage layer is required under footings if PWF footings are used.
- The bottom of the excavation must slope toward a drain.

If you decide to use preserved wood foundations, keep the following system considerations in mind.

System Considerations

Design

- The dimensions of the studs required and their spacing will depend on the backfill height. Higher levels of backfill may generate greater stresses on the wood, necessitating larger studs at more frequent intervals (refer to the CSA standard for more details).

Materials

- Lumber used in PWF systems must conform to CSA Standard 0322 PWF. Lumber suitable for use in preserved wood foundations will be stamped accordingly.
- If the lumber is cut in the field, cut surfaces must be thoroughly saturated with preservative by brushing or dipping.
- Follow the specific requirements for fasteners such as nails, staples, etc. and galvanized joist hangers and straps, etc. described in the CSA standard (see Figure 6.15). In many areas of the country, pre-ordering will be necessary.

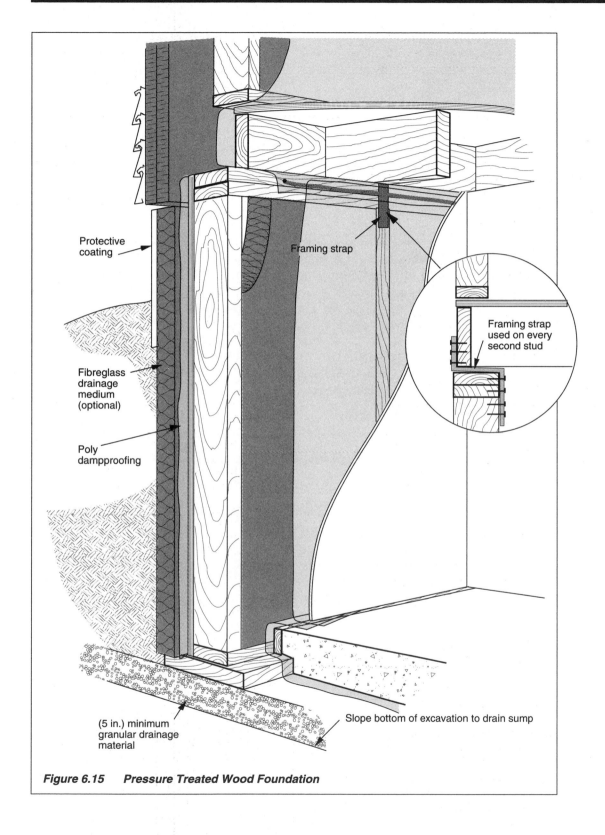

Protective coating

Framing strap

Framing strap used on every second stud

Fibreglass drainage medium (optional)

Poly dampproofing

(5 in.) minimum granular drainage material

Slope bottom of excavation to drain sump

Figure 6.15   Pressure Treated Wood Foundation

Labour and Supervision

- Site preparation may require additional supervisory time.
- Initial levelling for crushed gravel and wood footings must be true.
- Standard techniques can be used to frame and sheath walls, but framing straps must be used to connect the joist header to the foundation wall.

Air Sealing

- Concrete floors can be sealed by laying polyethylene sheeting under the slab and sealing it to the polyethylene sheeting in the walls.
- If plywood floors are used as the air barrier, all joints and seams must be caulked.
- If using sheet materials around the sill plate/floor joist assembly, sandwich the sheet between double top plates to prevent excessive wear during joist installation.

Dampproofing

- Use 0.15 mm (6 mil) polyethylene or commercial membranes for exterior dampproofing.
- Exterior joint sealants for the preserved plywood sheathing must be compatible with the sheathing and with the exterior dampproofing.
- To keep moisture from diffusing in from the soil, lay polyethylene dampproofing in strips over the crushed gravel drainage layer under the floor. The polyethylene should be lapped but not sealed to allow for the drainage of leaks.
- Conventional drainage tiles are not required. Lay a granular drainage layer over undisturbed soil to a depth of at least 125 mm (5 in.) and use a central gravity drained sump to provide drainage. If you are using concrete footings, pour the concrete over undisturbed soil and provide drainage holes in the footing.

6.6 UNHEATED CRAWLSPACES

An unheated crawlspace consists of a framed floor placed over a perimeter foundation wall and intermediate supports.

Conventionally, the crawlspace is vented all or part of the year. Recent studies have concluded that, providing moisture control procedures are put in place, venting of unheated crawlspaces may not be required.

Typically, insulation is placed between the floor joists and nested against the subfloor (see Figure 6.16). A variant, the suspended floor or open crawlspace, is used in arctic areas where it is necessary to prevent heat transmission from the building to the frozen soil (see Section 6.8). The techniques described in this section also apply to post-supported structures.

Unheated crawlspaces have several advantages.

- The crawlspace allows for high insulation levels.
- Good airtightness can be achieved.
- Excavation costs may be reduced.

However, there are also disadvantages to using this approach.

- Ductwork and water and sewer lines must be independently insulated or located on the warm side of the building envelope.
- The height of the space can restrict working conditions.
- Frost may damage the foundation under certain climatic and soil conditions.

Unheated crawlspaces have been unpopular in northern areas in the past because occupants complained of cold floors. This problem can be overcome by multi-layer construction and/or by circulating supply or return air between the subfloor and the finished floor.

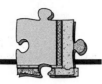

If you decide to use an unheated crawlspace, keep the following design considerations in mind.

- There must be an effective air seal.
- Convective air flow in the insulated cavities must be restricted.
- Thermal bridging in the insulation system must be minimized.
- The perimeter foundation wall or posts must have proper drainage and frost protection.
- Vapour diffusion from the soil must be prevented.
- Plumbing must be prevented from freezing. Ductwork must be carefully sealed and insulated.

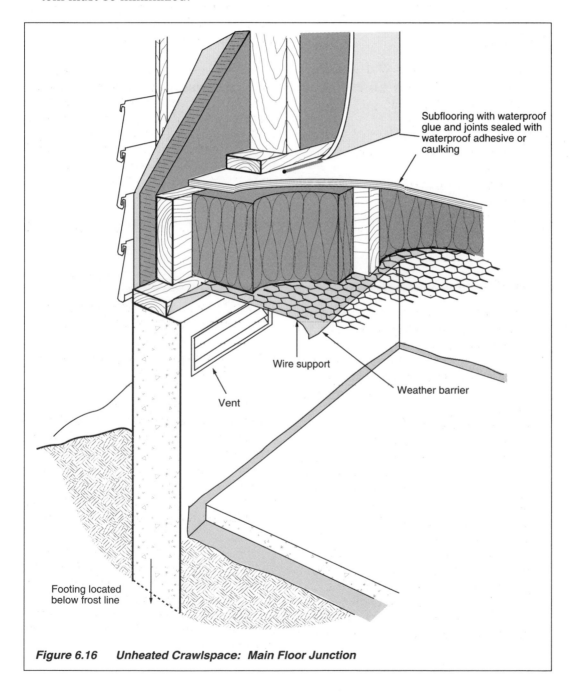

Subflooring with waterproof glue and joints sealed with waterproof adhesive or caulking

Wire support

Vent

Weather barrier

Footing located below frost line

Figure 6.16   Unheated Crawlspace: Main Floor Junction

Construction Sequence

A) Follow standard floor framing procedures but, after the roof and walls are made weather tight, either caulk all joints in the subfloor, glue them with a waterproof adhesive, or install 0.15 mm (6 mil) polyethylene, before installing the subfloor.

B) After the roof and walls are weather tight, cut holes wherever conduits or services penetrate the floor and install rubber gaskets. The holes should be slightly larger than the conduits to allow for their expansion and contraction due to temperature changes.

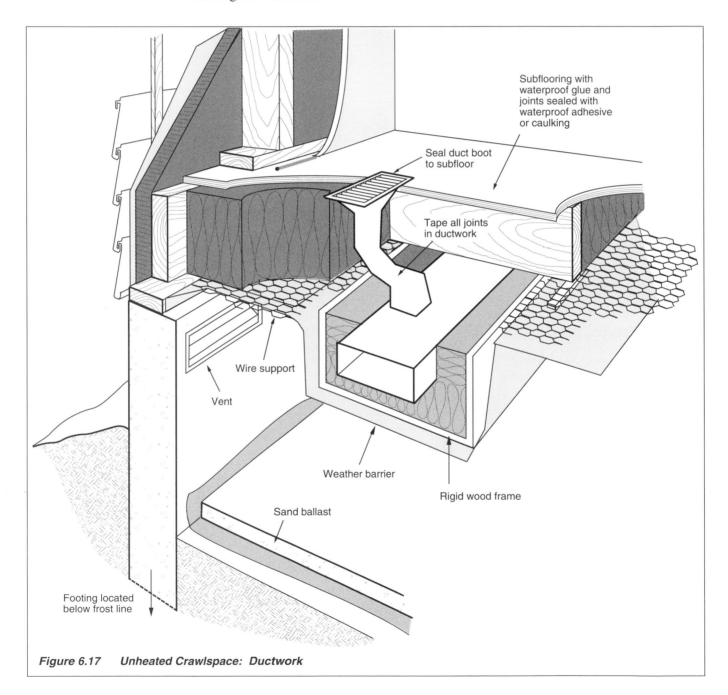

Figure 6.17   Unheated Crawlspace: Ductwork

C) Seal ductwork boots to the subfloor wherever they pass through it.

D) Seal ductwork with tape at all joints and along all seams.

E) After plumbing, ductwork and wiring are installed, place insulation between the floor joists in direct contact with the subfloor, supported below with a weather barrier such as a spun-bonded polyolefin or rigid sheathing material (see Figure 6.17).

F) Wrap all ductwork and plumbing with insulation of the same RSI (R) value as that used between the floor joists (see Figures 6.17 and 6.18).

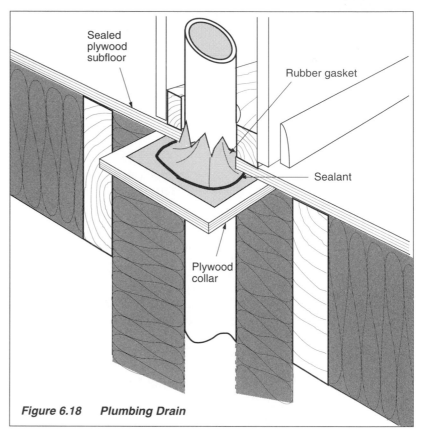

Figure 6.18 **Plumbing Drain**

6.7 SLAB-ON-GRADE

A slab-on-grade foundation consists of a slab with perimeter edge reinforcing, or a slab and a perimeter foundation wall. This type of foundation may be insulated beneath the slab and at the slab edge, or an insulated plywood floor may be placed over the slab.

There are several advantages to using slab-on-grade foundations.

- A slab-on-grade foundation may be cheaper to build than framed floors in some areas.
- A properly insulated slab can provide heat storage for solar and internal gains.
- A slab-on-grade foundation is less susceptible to moisture damage than an unheated crawlspace.

However, like other foundation approaches, slab-on-grade foundations also have disadvantages.

- Due to the higher cost of rigid insulations, slabs are generally not insulated to the same levels as unheated crawlspaces.
- Improper installation can lead to frost heaving.
- Slab-on-grade foundations are not suitable for areas with clay soils.

Slab-on-grade construction must be designed to resist frost heave. This design includes protecting footings from freezing and properly draining surrounding soils (see Figure 6.19). These foundations are commonly insulated below the slab with rigid board insulation. Several points should be kept in mind.

- The largest area of heat loss occurs at the slab edge. Insulate to levels equal to the exterior walls, or as high as practical. In some cases, a thermal break of rigid insulation is placed between the slab edge and the footing.

- Isolate the floor slab from the soil and outdoor air temperatures with insulation.

- Insulate foundation walls from the exterior to reduce the chance of frost penetration.

- Pour the slab on a 125 mm (5 in.) bed of compacted gravel, which can be covered with high compressive strength rigid insulation.

- A polyethylene moisture barrier may be installed beneath the slab. To prevent the slab from cracking and to protect the polyethylene when pouring the floor, some builders cover the polyethylene with 50 mm (2 in.) of sand or with rigid insulation.

- In most parts of the country, RSI 1.8 (R-10) of rigid board insulation should be installed under the slab. Higher levels may be necessary depending on climate, soil and ground water conditions. In frost susceptible soils, the footing should be further protected by an insulating skirt (see (A) in Figure 6.19).

- Heat loss from the underside of the slab is high, particularly at the edge. Run a 0.6 m (2 ft.) wide strip of insulation around the entire perimeter of the slab (see (B) in Figure 6.19). A vertical band of rigid insulation in the exterior wall will also cut heat loss and reduce the chance of frost penetration under the slab (see (C) in Figure 6.19).

- Insulation and strapping above the slab will provide a warmer, more flexible and hence more comfortable floor surface.

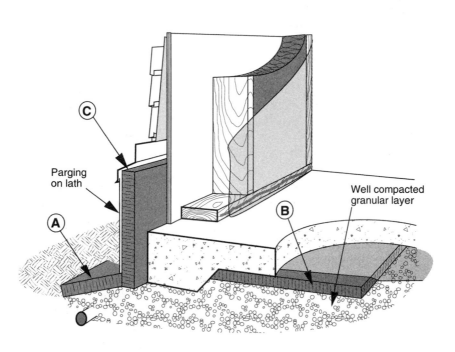

Slab-on-grade Membrane Air Barrier Approach

Labels in upper figure: C, Parging on lath, A, Well compacted granular layer, B

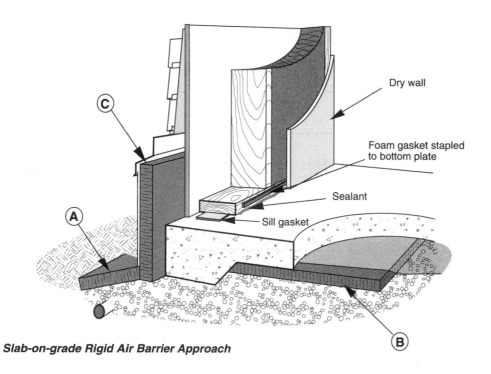

Slab-on-grade Rigid Air Barrier Approach

Labels in lower figure: C, A, Dry wall, Foam gasket stapled to bottom plate, Sealant, Sill gasket, B

Figure 6.19   Slab-on-grade

6.8 NORTHERN FOUNDATIONS

Foundation design in northern areas presents special problems, partly because of difficulties in access, availability of materials and timing, but mainly because of the difficult environmental and soil conditions. Be sure to study local practices and consult local experts before designing a foundation for a northern house.

6.8.1 Soil Conditions

In the northern parts of many provinces there are large areas of muskeg or soils with high water tables. Other areas with deeper water tables may have frost susceptible soils and/or conditions that encourage ice lensing. These conditions, combined with deep frost penetration, make it difficult to build stable foundations.

Conditions are even more challenging further north where there are areas of discontinuous or continuous permafrost.

Permafrost is soil with a high water content that remains below 0°C (32°F) throughout a year. It can occur in scattered patches surrounded by soil that experiences normal freeze-thaw cycles, but farther north it occurs as large continuous areas of frozen soil, several hundred metres thick.

In all cases, permafrost is covered by a layer of soil that experiences normal freezing and thawing through the seasons. This "active layer" can vary in thickness from a few centimetres in high arctic areas to several metres.

Soils in the active layer may be partially or fully saturated. If they are frost susceptible, considerable movement may be expected in this layer.

	Soils Not Susceptible to Frost	**Frost Susceptible Soils**
Nonpermafrost	• Use normal foundation techniques. • Shallow footings are possible.	• Techniques used are similar to those used for southern foundations. • Use deeper footings to reach below the frostline. • Consider clear span joists in crawlspaces to reduce differential settlement. • Use a basement only if problems of water or moisture penetration can be handled.
Discontinuous Permafrost	• The approach recommended for permafrost with frost susceptible soils may be necessary. • Carry out a detailed soils study and consult local expertise before deciding on an approach.	
Permafrost beneath the "active layer" of soil	**High water table** • Minimize heat transmission from the building through the active layer to the permafrost. • Basements and heated crawlspaces are not advisable. • Open or ventilated crawlspaces are the best solution. Place a gravel pad on the surface to provide a stable and well drained top layer. **Water table well below the top of the permafrost** • Footings and foundation practices used in southern Canada may be used. • Consult local practice as well.	• Minimize heat transmission from the building through the active layer to the permafrost. • Basements and heated crawlspaces are not advisable. • Open or ventilated crawlspaces are the best solution. Place a gravel pad on the surface to provide a stable and well drained top layer.

Figure 6.20 Approaches to Foundation Design for Northern Soil Conditions

It is essential to avoid thawing the permafrost, which will produce an unstable soil.

Different soil conditions require different approaches to foundation design (see Figure 6.20).

6.8.2   Northern Foundation Types

A number of foundation systems have been developed specifically for northern use. In post and pad or pad and wedge systems, footings can be located on or in the pad with the expectation of some movement and future adjustment. Other systems such as piles or Greenland foundations can be taken right to the more stable permafrost layer. These systems are more permanent and provide better anchorage which is crucial in high wind areas.

Piles

Piles are suited to sites where the soil has low bearing capacity. Since the small cross-sectional area of piles minimizes heat transfer from the building to the soil, piles are especially suitable for permafrost applications. The penetration of piles into the frozen permafrost layer ensures stable support, even if the soil in the active layer has low bearing capacities.

Piling also provides anchorage against uplift, an important consideration in areas with high wind speeds. Although piles can be made of wood (see Figure 6.21) or concrete, steel piles are becoming the standard because of their greater durability (see Figure 6.22). Piles in permafrost are usually placed in a predrilled hole and backfilled with a sand/water slurry which, freezes, providing a firm grip on the pile.

The lifting forces that result from annual freeze / thaw cycles may cause problems. These problems can be overcome by using drill rod grease and properly anchoring the piles in the permafrost. Piles should be designed by an engineer familiar with local conditions.

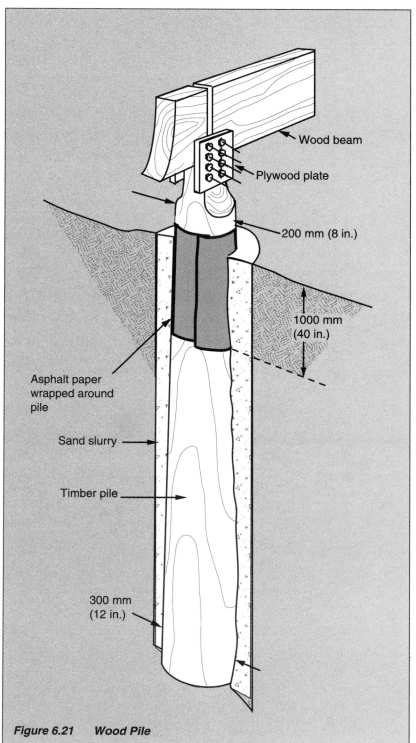

Figure 6.21   Wood Pile

Labels in figure:
- Wood beam
- Plywood plate
- 200 mm (8 in.)
- 1000 mm (40 in.)
- Asphalt paper wrapped around pile
- Sand slurry
- Timber pile
- 300 mm (12 in.)

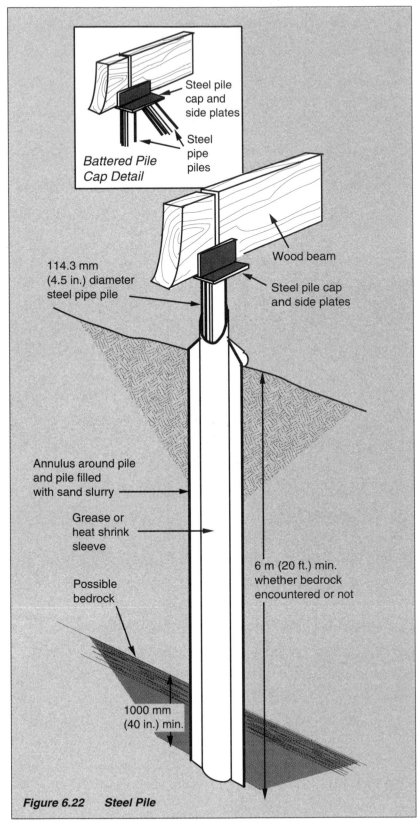

Battered Pile Cap Detail

Steel pile cap and side plates

Steel pipe piles

Wood beam

114.3 mm (4.5 in.) diameter steel pipe pile

Steel pile cap and side plates

Annulus around pile and pile filled with sand slurry

Grease or heat shrink sleeve

Possible bedrock

6 m (20 ft.) min. whether bedrock encountered or not

1000 mm (40 in.) min.

Figure 6.22   Steel Pile

Skirting is often used in conjunction with piles and pilasters to create a crawlspace. However, this skirting must not restrict the passage of wind in areas where drifting snow is a problem. The skirting must also not promote thawing of the permafrost below the house; it must be at least 300 mm (12 in.) above the ground if you think thawing may be a problem.

Grade Beams

This system is less commonly used. It involves laying concrete strip footings or laminated wood grade beams on top of a prepared gravel pad. The system is relatively expensive because it relies on concrete or laminated wood, and is susceptible to uneven settlement unless the pad is very stable.

Greenland Foundation

This system consists of a wood pad sitting on undisturbed and frozen permafrost (see Figure 6.23). The construction process involves rapid excavation of a hole to a level below the active layer, rapid placement of the pad and post, and quick backfilling to prevent thawing of the exposed frozen soil. Site supervision must be good to ensure that work is completed before the permafrost melts. The economics of this approach are questionable.

Crib or Drum

In this system, wood cribs or drums are set into excavated holes on prepared gravel pads and filled with coarse gravel or rocks (see Figure 6.24). This is a more permanent solution than the pad and wedge method, and is a good way to underpin and stabilize existing pad and wedge systems that have been undermined. The system requires undisturbed permafrost. Use preserved wood for the cribs, especially in areas where the ground thaws.

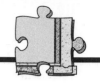

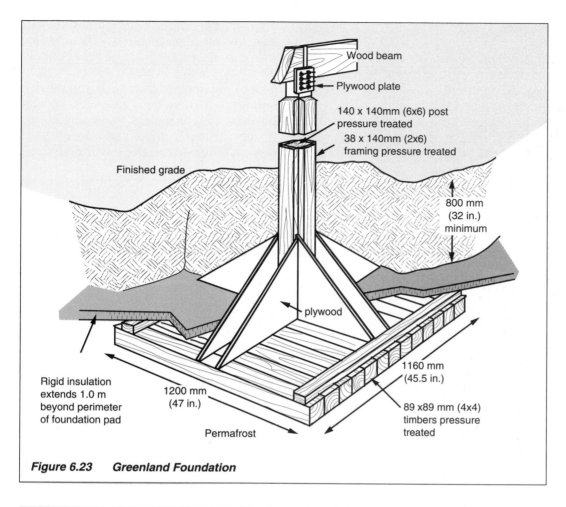

Wood beam

Plywood plate

140 x 140mm (6x6) post
pressure treated

38 x 140mm (2x6)
framing pressure treated

Finished grade

800 mm
(32 in.)
minimum

plywood

1160 mm
(45.5 in.)

Rigid insulation
extends 1.0 m
beyond perimeter
of foundation pad

1200 mm
(47 in.)

89 x89 mm (4x4)
timbers pressure
treated

Permafrost

Figure 6.23   Greenland Foundation

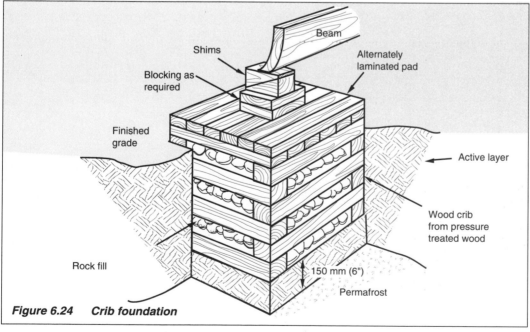

Beam

Shims

Blocking as
required

Alternately
laminated pad

Finished
grade

Active layer

Wood crib
from pressure
treated wood

Rock fill

150 mm (6")

Permafrost

Figure 6.24   Crib foundation

Pad and Wedge

This system consists of timbers laid in alternating layers on a prepared gravel pad or exposed rock outcrop (see Figure 6.25). Use preserved wood in all but high arctic areas. Since a large volume of wood is required, the system is only appropriate in areas where material can be shipped by road or barge.

This system can be used with soils that have low bearing capacities. It is suited for use in small buildings which are not likely to suffer damage from a considerable degree of movement, and is less suitable for buildings with wood frame and drywall finishes. Nevertheless, it is a standard system in arctic areas and can be used successfully with proper levelling.

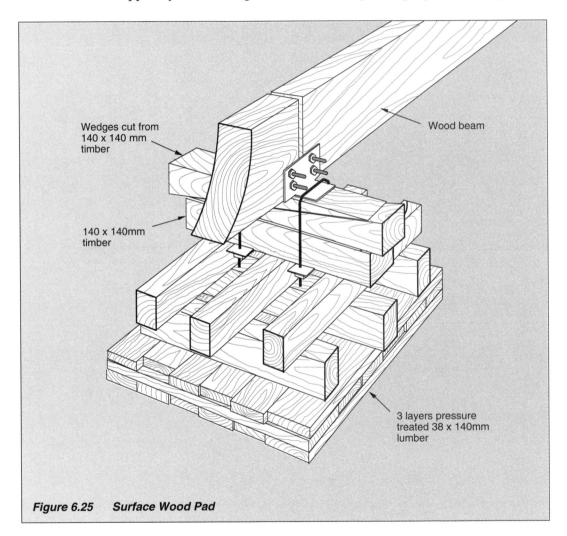

Wedges cut from 140 x 140 mm timber

140 x 140mm timber

Wood beam

3 layers pressure treated 38 x 140mm lumber

Figure 6.25   Surface Wood Pad

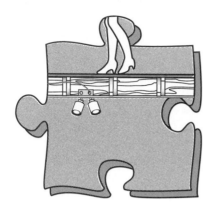

Chapter 7
FLOORS

From the point of view of maintaining comfortable conditions and preventing cold floors, the intersections of floors and exterior walls are critical. Special attention must be paid to these intersections in order to maintain proper insulation levels and continuity of the air barrier. Doing the job correctly during construction can reduce the much greater cost of remedial measures after completion.

This chapter looks at different approaches to dealing with main floors, intermediate floors, cantilever floors and floors over un-heated crawlspaces and garages. Sections 7.5 through 7.7 look at the special requirements of floors in multi-unit buildings, techniques to be used with floor trusses, and modified balloon framing options.

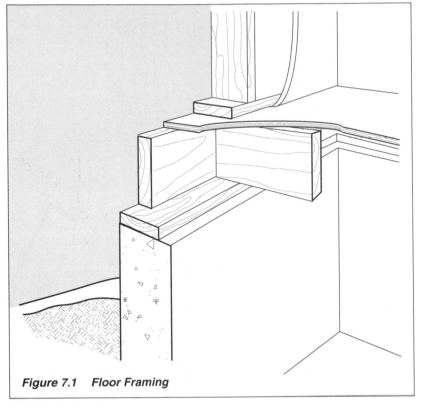

Figure 7.1   Floor Framing

7.1   *MAIN FLOOR ASSEMBLY OPTIONS*

The main floor is generally considered to be the floor assembly that rests on the foundation wall. It is usually framed in one of two ways.

In the first method (see Figure 7.1), the sill plate is levelled on top of the foundation wall. The header or band joist and all floor joists are installed and then the subfloor is applied. The second method (see Figure 7.2) represents an alternative. The header or band joist and the floor joists are embedded in the concrete when the foundation walls are poured.

No matter which method is used, the header or band joist area must be insulated and air leakage minimized if the building is to perform as expected. Sill gasket materials can reduce the potential for air leakage between the framing materials and concrete. Either rigid or membrane air barriers may be used.

Figure 7.2   Floor Framing — Western Canada Method

7.1.1 Membrane Air Barrier Approach

Figure 7.3 illustrates a technique that can be used to seal and insulate the header joist area if polyethylene is being used as both the air barrier and the vapour diffusion retarder. The header joist is recessed and the polyethylene is passed under and around it during assembly. Note that the exterior insulation must have at least twice the thermal resistance of any insulation placed inside the header.

Alternatively, frame the detail without recessing the header and apply batt-type insulation inside the header (Figure 7.4). Then cover it with an air barrier cut to fit between the joists. Seal the air barrier to both sides of the floor joist, to the subfloor, and to the air barrier system on the foundation wall. All cracks and joints in the subfloor which are outside of the air barrier must be sealed. This approach is not recommended if plank subflooring is used.

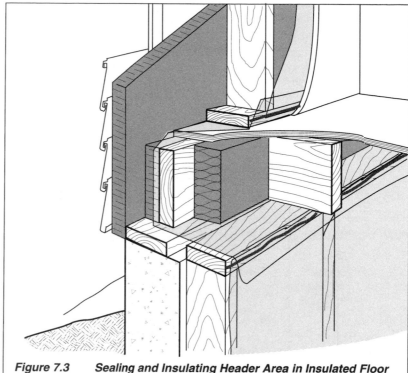

Figure 7.3   Sealing and Insulating Header Area in Insulated Floor Assembly (Membrane Air Barrier Approach)

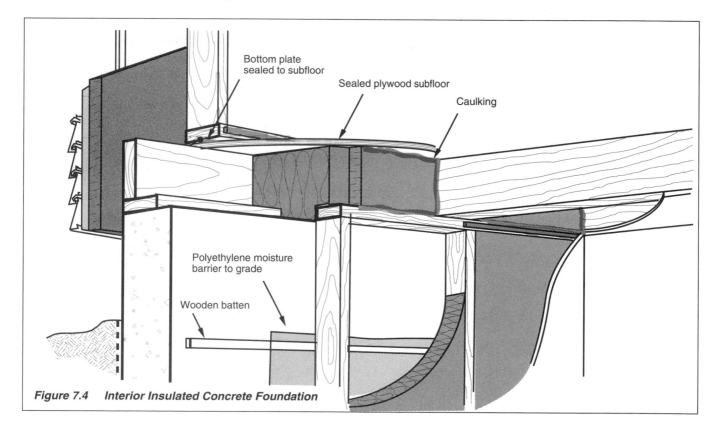

Bottom plate sealed to subfloor

Sealed plywood subfloor

Caulking

Polyethylene moisture barrier to grade

Wooden batten

Figure 7.4   Interior Insulated Concrete Foundation

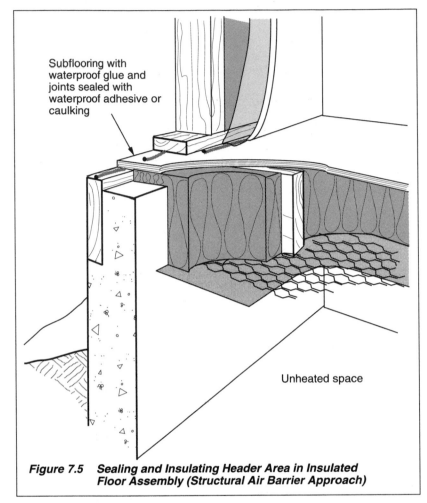

Subflooring with waterproof glue and joints sealed with waterproof adhesive or caulking

Unheated space

Figure 7.5 *Sealing and Insulating Header Area in Insulated Floor Assembly (Structural Air Barrier Approach)*

7.1.2   Rigid Air Barrier Approach

In this approach, the wood members and panel products such as subflooring are used as part of the air barrier. Gasketing and/or sealing is applied between the plates, the header joist, and the subfloor. Vertical joints in the header joist must also be sealed.

If the floor joists are embedded in the concrete, install gasket material or a sealant between the top edge of the embedded header joist and the underside of the subfloor sheathing. Then, seal the bottom edge of the embedded header joist to the cast concrete wall with an appropriate sealant or caulking material (see Figure 7.5). All butt joints between sections of the header joist must also be sealed.

If the subfloor is being glued in addition to being screwed or nailed to the floor joists, a continuous bead of the glue instead of the caulking or gasket is applied between the header joist and the subfloor. The gasket under the bottom plates of the exterior perimeter walls may also be replaced with a bead of caulking at the inside edge of the bottom plate where it meets the subfloor. This bead of caulking is normally installed along with the other gaskets and sealants after the framing is complete and before the gypsum board is installed.

Set the joist toward the inside of the wall plate and insulate on the cold side. This will shift the dew point enough to limit moisture formation on the interior surface of the header joist.

7.2 INTERMEDIATE FLOOR ASSEMBLY OPTIONS

7.2.1 Membrane Air Barrier Approach

Figure 7.6 shows a method that can be used to frame intermediate floors while maintaining continuity of the air barrier. Sandwich a 600 mm (24 in.) strip of polyethylene between the top plates during wall assembly and drape it to the outside before the floors are constructed. After the subfloor is laid, wrap the strip around the subfloor wood panel sheathing before the next wall is framed in place. Install rigid board insulation over the header in the space provided by recessing the joists 38 mm (1-1/2 in.) from the exterior of the studs.

An optional wood panel sheathing nailer may be installed before the exterior sheathing is nailed in place. This may make siding installation easier, eliminating the need for 100 mm (4 in.) siding nails. Batt insulation representing no more than one-third of the thermal resistance value of this strip can be placed in the interior joist cavities. The technique described in Figure 7.4 can also be used in this location.

If electrical wiring or the plumbing is run through top or bottom plates, seal the penetrations with caulking.

7.2.2 Rigid Air Barrier Approach

Figure 7.7 shows how to maintain air seals at subfloors between storeys when using a rigid air barrier. Drywall must be sealed to the top and bottom plates. Where floor joists run parallel to the exterior perimeter walls, the details shown in Figure 7.8 can be used to maintain the continuity of the air barrier. Where a ceiling drop is required, the details shown in Figure 7.9 may be used.

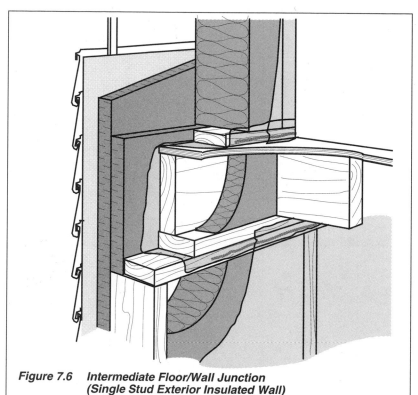

Figure 7.6   Intermediate Floor/Wall Junction (Single Stud Exterior Insulated Wall)

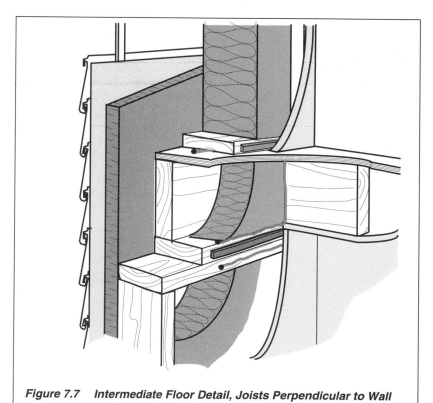

Figure 7.7   Intermediate Floor Detail, Joists Perpendicular to Wall

107

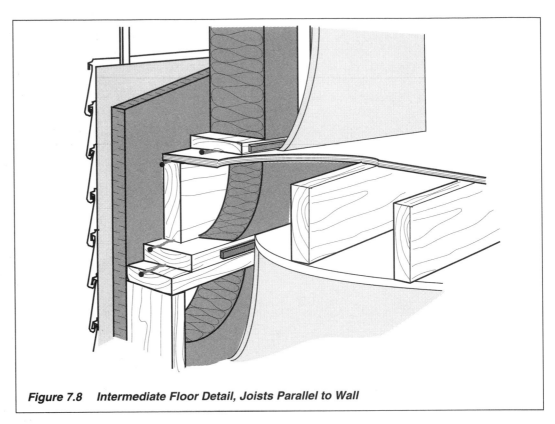

Figure 7.8   Intermediate Floor Detail, Joists Parallel to Wall

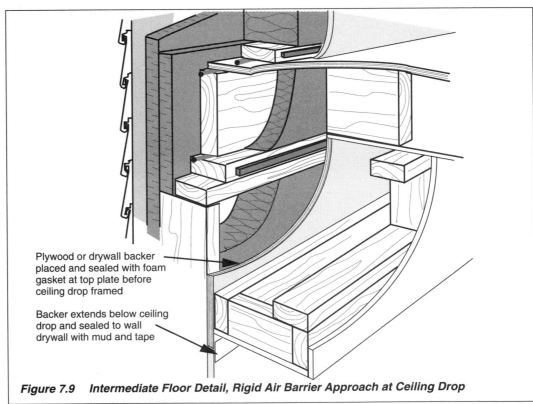

Plywood or drywall backer placed and sealed with foam gasket at top plate before ceiling drop framed

Backer extends below ceiling drop and sealed to wall drywall with mud and tape

Figure 7.9   Intermediate Floor Detail, Rigid Air Barrier Approach at Ceiling Drop

7.3 CANTILEVER FLOORS

If a sheet of polyethylene is not laid under the subfloor during construction of a cantilever floor, then the seams in the subfloor must be sealed to serve as the air barrier. Before insulating the cantilever, ensure that there is no water trapped between the subfloor and the polyethylene. Insulation should be placed to fill the entire cavity, from the header joist to the inside face of the wall below (see Figure 7.10a).

Blocks of wood or rigid polystyrene insulation can be installed between the floor joists. The edges of the blocking must be sealed to the subfloor above, the floor joists on either side, and the air barrier from the wall below.

The polyethylene from the wall above is sealed to the subfloor. The underside of the overhang must be protected by a weather barrier connected to the weather barriers of the walls above and below.

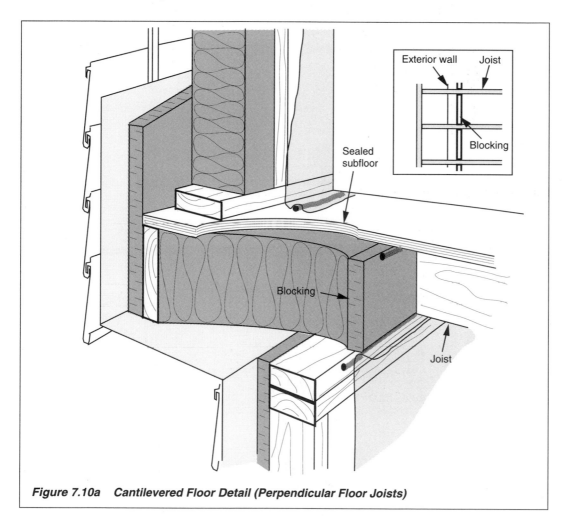

Figure 7.10a   Cantilevered Floor Detail (Perpendicular Floor Joists)

If the floor joists run parallel to the exterior wall over which a cantilever will extend (see Figure 7.10b), the cantilever must be constructed using "ladders" in the same way a roof overhang is built to pass over the gable end of a house.

The techniques described above can be applied in this case as well. Where cantilevers support roof loads, they must extend into the framing six (6) times the length of the cantilever.

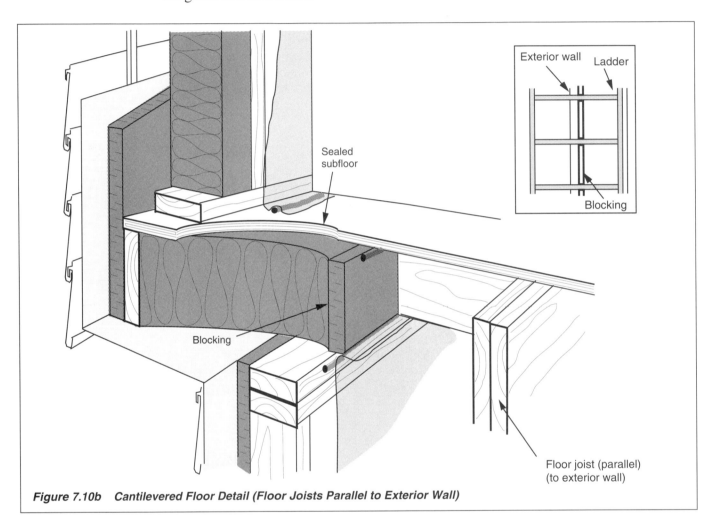

Figure 7.10b   Cantilevered Floor Detail (Floor Joists Parallel to Exterior Wall)

7.4 FLOORS OVER UNHEATED CRAWLSPACES AND GARAGES

Floors over unheated crawlspaces and garages must be insulated, and the joint between the header or band joist and the subfloor must be sealed to keep cold air from short-circuiting the insulation. The joint between the top plate and the header joist must be sealed, as must all air leakage sites in the floor. These include plumbing and electrical penetrations, and all openings where heating ducts pass through the floor into the space above.

If polyethylene is not installed under the subfloor, it will also be necessary to seal all of the seams in the subfloor. Ductwork must be sealed along all seams and at each joint.

The insulation should be installed snugly to the subfloor after the services have been installed (see Figure 7.11). Ductwork should be insulated to the same level as the floor (see Figure 7.12). The insulation should be protected by a weather barrier to keep wind out of the insulated space.

In garages, the insulation should be protected by drywall, for fire protection and to keep gases out of the dwelling unit. Spray-in-place foam insulation has been effectively applied in this location, reducing air leakage while providing high insulation levels.

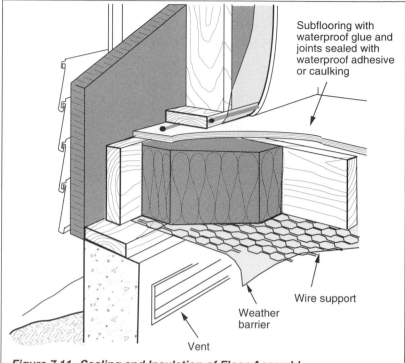

Figure 7.11 Sealing and Insulation of Floor Assembly (Rigid Air Barrier Approach)

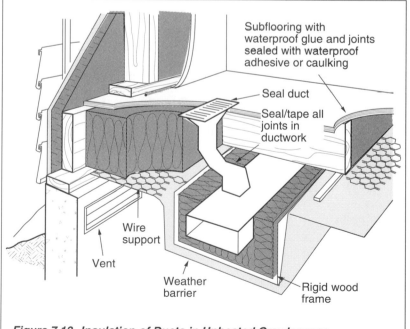

Figure 7.12 Insulation of Ducts in Unheated Crawlspaces

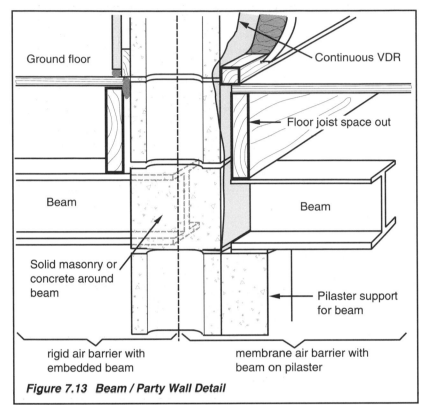

Ground floor

Continuous VDR

Floor joist space out

Beam

Beam

Solid masonry or concrete around beam

Pilaster support for beam

rigid air barrier with embedded beam

membrane air barrier with beam on pilaster

Figure 7.13   Beam / Party Wall Detail

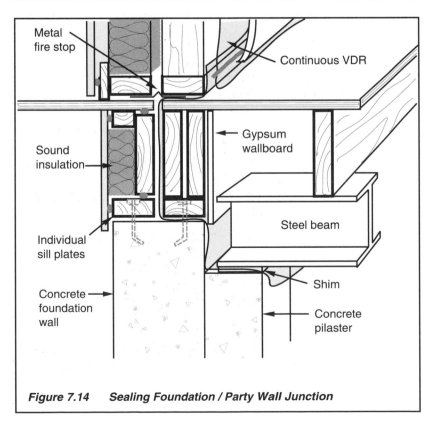

Metal fire stop

Continuous VDR

Sound insulation

Gypsum wallboard

Individual sill plates

Steel beam

Concrete foundation wall

Shim

Concrete pilaster

Figure 7.14   Sealing Foundation / Party Wall Junction

7.5   FLOORS IN MULTI-UNIT BUILDINGS

Floors in multi-unit buildings require special attention in order to reduce sound transmission from one unit to another and to lessen the potential of fire spreading between units.

7.5.1   Beam/Party Wall Details

The main floor can be framed in a fashion similar to that used for detached houses; the same structural requirements apply (see Figure 7.13). Beam ends bearing on the party wall should be separated by 100 mm (4 in.) of solid masonry to minimize sound transfer between units, and all gaps in the masonry should be sealed. Consideration should be given to supporting the beams on pilasters rather than setting them in the foundation wall.

Air leakage up the party wall from the basement can be eliminated by building the floor assembly as shown in Figure 7.14. The left half of the illustration shows the rigid air barrier approach and the right half illustrates the membrane air barrier approach. Note how the subfloor is discontinuous and separate bottom plates are used. This procedure helps to reduce sound transmission from one unit to another.

7.5.2 Floor/Party Wall Details

If the floor joists run perpendicular to the party wall and are supported by masonry, they must rest on a solid bearing support (see Figure 7.15 left side). The joist ends must be fire cut so that, in the event of a fire, collapsing joists will not put upward and outward pressure on the masonry wall.

The party wall between the joist ends should be at least 100 mm (4 in.) thick, and all gaps around the joists must be sealed. These measures will help to minimize sound transfer and smoke movement between units. Sound transmission will be further reduced if the floor joists run parallel to the party wall and the subfloor is kept away from the wall. Care must be taken to eliminate air leakage at this location. Techniques for sealing the bottom plate/subfloor joint are described in Chapter 8.

Figure 7.16 illustrates another approach to framing floors where they connect to party walls. A strip of polyethylene is placed so that it can be folded around the header after the upper floor has been framed (see the right half of the illustration). The vertical air space should also be blocked by a metal fire stop. The left half of the illustration shows how the detail can be framed using a rigid air barrier. Note the sound insulation placed in the header area.

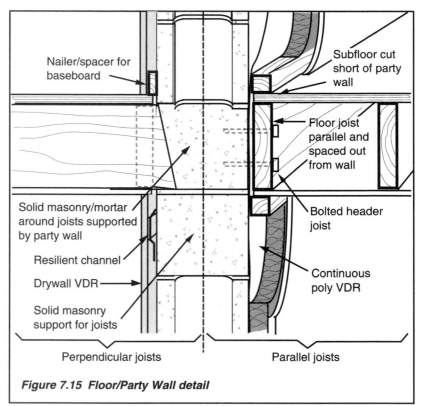

Figure 7.15 Floor/Party Wall detail

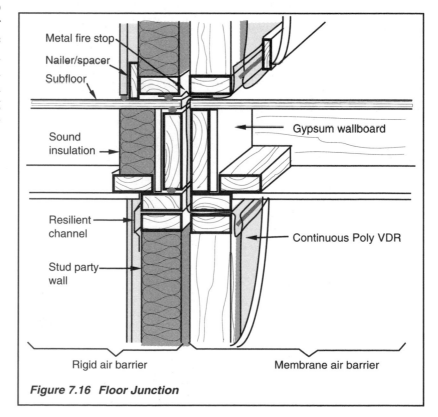

Figure 7.16 Floor Junction

7.6 USE OF FLOOR TRUSSES

Many builders are using floor trusses rather than standard floor joists. Floor trusses provide the benefits of allowing for increased spans without increasing the depth of the floor assembly. Floor trusses are also less subject to the shrinkage and twisting that result from the higher moisture content of conventional framing materials.

In addition to spanning greater distances, trusses can use smaller-dimension lumber. They are also becoming increasingly price-competitive with floor joists. Trusses can be designed to accommodate heating and ventilating ducts so that they do not reduce head room in the storey below. Electrical wiring and plumbing can be run between the top and bottom chords without requiring drilling.

Trusses, unlike standard joists, do not shrink. Consequently, floors built with trusses are less likely to squeak. When using trusses, rather than using a standard wood joist, you should use wood panel sheathing as the header or band joist because it too does not shrink (see Figure 7.17).

For a membrane air barrier, use the techniques described in Section 7.1.1. For a rigid air barrier, use a construction adhesive to seal between the top of the wood panel sheathing header and the wood panel sheathing subfloor because it is usually too difficult to install a gasket. At the bottom of the wood panel sheathing header, where it may be difficult to position a narrow gasket, a wide gasket with a rectangular profile may be used. Any sealant used must allow for building frame movement.

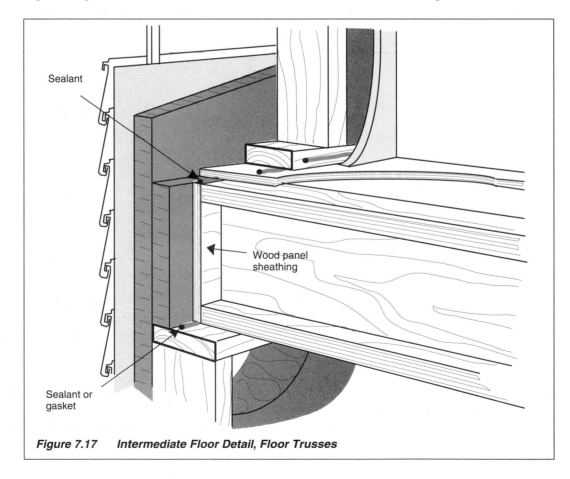

Sealant

Wood panel
sheathing

Sealant or
gasket

Figure 7.17   Intermediate Floor Detail, Floor Trusses

7.7 MODIFIED BALLOON FRAMING DETAILS

Prior to World War II, many of the houses built in Canada were constructed using standard balloon framing techniques (see Figure 7.18). In a balloon-framed building, the wall studs are continuous from the sill plate up to the double top plate, even in a two-storey house.

Balloon framing can have advantages. Most notably, it provides for complete insulation of the wall, except for the thermal bridges created by studs and plates, without the problems caused by the header joists in platform framing. However, because the floor joist ends extend into the wall cavity, special care is needed to achieve a good air seal at these locations. This section discusses modified balloon framing techniques that can be used to ensure good air sealing at wall/floor junctions.

7.7.1 Foundation Wall/Main Floor Detail

A modified balloon framing technique can be used to ensure continuity of the air barrier as well as full depth insulation at the basement wall/main floor junction (see Figure 7.19). A 38x190 mm (2x8 in.) sill plate is installed on top of the foundation wall. Prior to floor assembly installation, the polyethylene air barrier/vapour diffusion retarder is laid under the plate; a strip slightly wider than 1.2 m (48 in.) is needed so that 0.6 m (24 in.) will overhang each side of the wall. The polyethylene should be protected with building paper during assembly of the floor and wall systems.

The floor system is then installed as in a platform-framed system, except that it is kept back 100 mm (4 in.) from the outer edges of the plates. The subfloor is installed, and the polyethylene is brought up to wrap the header and lie on top of the subfloor.

Figure 7.18   Standard Balloon Framing

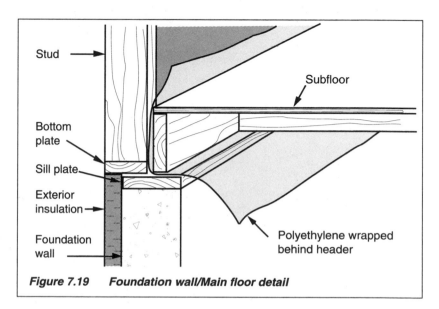

Figure 7.19   Foundation wall/Main floor detail

The wall is framed on the deck and set down onto the sill plate outside the floor assembly. Blocking or strapping must be installed between the studs to provide fastening points for the polyethylene, drywall and baseboard.

Insulation can be installed outside the header joist between the studs. Insulating sheathing can also be installed on the exterior to provide a higher RSI (R) value for the system.

7.7.2 Intermediate Floor Detail

Intermediate floors can also be framed using a modified balloon framing technique (see Figure 7.20). The first-floor walls are framed as described in Section 7.7.1. A strip of polyethylene is installed, leaving enough hanging over the top plate to seal to the polyethylene that will be applied to the second-storey walls.

The polyethylene strip should also be wide enough to hang down below a ledger strip made of lumber with the same dimensions as the floor joists. This ledger strip is installed so that its top edge is flush with the top edge of the single top plate of the first-storey walls. The floor joists are then installed using joist hangers.

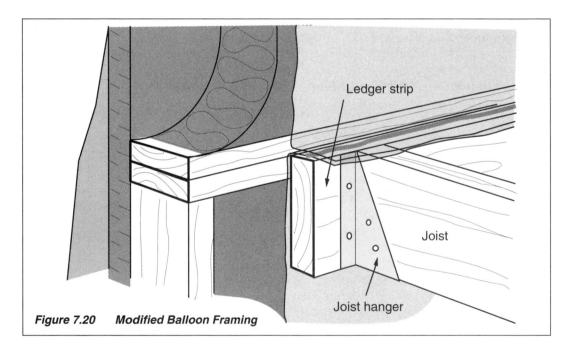

Figure 7.20   Modified Balloon Framing

Chapter *8*

WALLS

In addition to supporting upper floors and the roof, exterior walls must provide a visual, thermal, weather, humidity, noise and dust barrier between the interior and exterior. To perform these functions properly, an exterior wall must:

- resist external forces such as wind, rain, snow and sun
- resist the outward migration of both conditioned air and water vapour, and
- resist heat flow in both directions.

To meet these requirements, modern walls contain at least the following components:

- external skin or cladding
- a weather barrier
- structural components including framing members, sheathing, etc.
- insulation
- an air barrier
- a vapour diffusion retarder, and
- interior skin or finish.

In a properly constructed wall, these components work together to produce a durable, structurally sound system that is resistant to heat, air and moisture flows. This chapter outlines different approaches to constructing insulated exterior walls with effective barriers to air and moisture flow.

8.1 SELECTING A WALL SYSTEM

When deciding what kind of wall system to use, you should consider:

- local climatic conditions
- the level of thermal performance required
- market acceptance of different systems
- materials availability and cost, and
- labour and supervisory requirements.

Sections 8.4 through 8.9 describe different approaches to building structurally sound insulated wall systems. First, however, some of the principles behind the different construction approaches are discussed in Sections 8.2 and 8.3.

8.2 MINIMIZING HEAT LOSS

Advanced framing techniques such as those described in this chapter can reduce the amount of lumber used in framing walls. Since lumber has a very low RSI value, this can reduce heat loss through walls by up to 10 percent, if the techniques are applied consistently throughout the house. Some of the techniques described increase wall thickness, which allows for more insulation. Whatever technique is used, the insulation must be installed properly to obtain the full benefit. It must fill the entire cavity, leaving no gaps where convection currents could form. In double stud and standoff walls, different batt thicknesses may be required in different parts of the wall to completely fill the cavity. Walls must also be constructed to prevent cold air from circulating through or around the insulation.

Over the past few years, various blown and spray insulations have been developed. Spray foam insulations and wet and dry cellulose and fibreglass systems, when applied correctly, can fill cavities and control convection. Some spray insulations are also effective in reducing air movement through the wall assembly.

Several manufacturers produce building assemblies that incorporate framing members into solid foam sections (see Sections 8.8 and 8.9). These systems address some of the shortcomings of improperly installed batt insulation if joints are sealed. They are, however, not as flexible as "stick built" systems. The manufacturers of the various products provide training on how to use their systems.

Heat loss through framing members is considerably greater than through the insulation. Reducing framing lumber can reduce energy losses and lower labour and material costs. Techniques that you should consider include:

- using 38x140 mm (2x6 in.) framing lumber at 600 mm (24 in.) on centre and alternative wall board techniques to minimize deflection
- lining up floor and ceiling joists over studs to eliminate the need for double top plates
- using drywall clips at corners to eliminate three stud corners and reduce cracks in drywall due to wood shrinkage and truss uplift (see Figure 8.1)
- using horizontal blocking and drywall clips where partition walls intersect exterior walls (see Figure 8.2)
- using modified balloon framing techniques to reduce wood shrinkage and thermal bridging through top and bottom plates (see Figure 8.3), and
- using diagonal bracing in conjunction with insulating sheathing to eliminate the need for structural sheathing and to reduce thermal bridging through wall systems.

These techniques can be used to build any of the wall systems described in Sections 8.4 through 8.7.

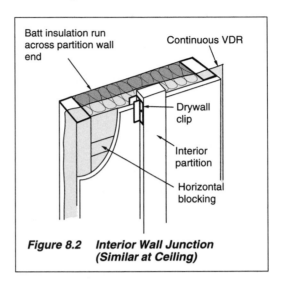

Figure 8.2   Interior Wall Junction (Similar at Ceiling)

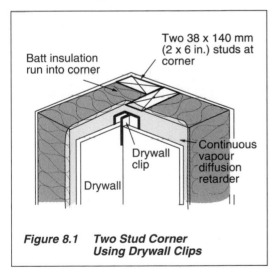

Figure 8.1   Two Stud Corner Using Drywall Clips

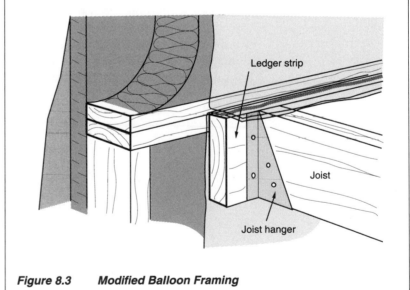

Figure 8.3   Modified Balloon Framing

8.3 MINIMIZING MOISTURE PROBLEMS

8.3.1 Moisture Considerations with Exterior Sheathing

A wall system can get wet because water enters from the inside or the outside, or is released from wood members used to build the house. Several factors influence the rate at which it will dry:

- the permeability of the exterior sheathing/cladding system
- the number and location of joints in the exterior sheathing/cladding system
- the length of the drying season, and
- the relative humidity of the outside air during the drying season.

Moisture problems in walls sheathed or clad with low-permeance sheet materials have been a particular concern for many years. Moisture entering the walls from inside can be trapped behind the sheathing or cladding, leading to structural damage. As far back as 1953, the National Building Code of Canada stated that a Type I vapour diffusion retarder must be installed on the high vapour pressure (warm) side of any walls sheathed with low-permeance materials.

Studies have been undertaken to determine the nature and extent of the problem in existing houses, and to assess the drying capacity of different wall systems. Test panels studied have included:

- walls with waferboard sheathing, both with and without exterior strapping
- walls with glass fibre sheathing, both with and without exterior strapping
- walls with extruded polystyrene sheathing, both with and without exterior strapping
- walls incorporating expanded polystyrene sheathing, and
- walls incorporating spray-applied cellulose insulation.

Researchers working on one major study made the following observations.

- The framing lumber purchased at local outlets had moisture contents of 26 to 30 percent, which is above the fibre saturation point of wood (19 percent). All of the test panels started to dry during the monitoring phase of the study.

- South-facing panels dried more readily than north-facing panels of the same construction.

- Panels with permeable sheathing systems dried more readily than those sheathed in low-permeance materials.

- Most panels which stay wet for prolonged periods exhibited some fungal growth on wood-based sheathing materials and framing lumber.

- Creating a rainscreen with strapping may help to reduce the accumulation of moisture in wood framing members and wood-based sidings and sheathing materials.

- Water penetration from the outside was a frequent contributor to moisture problems in walls.

These results did not indicate that any one system performs better than the others. However, the observation about the importance of water penetration from the outside lends support to the use of "rainscreens" to help reduce the potential for moisture problems in walls.

8.3.2 The "Rainscreen" Principle

The frequency and quantity of water penetration into a wall from the outside depends on:

- air leakage through the exterior cladding
- variations in air pressure inside and outside the wall
- weather conditions, particularly rain and wind, and
- the amount of solar radiation — sunlight heating the wall after a rainfall, for example, can vaporize water running down the back of brick cladding; under certain conditions, this water vapour may then be pushed into the wall.

Water will only penetrate if there is a hole for it to enter and a pressure difference to push it through. Cladding materials generally offer an abundance of cracks and holes for water to leak through. The pressure pushing the water through may be due to wind, capillary action, gravity as when water runs down the back of bricks, and/or differences in the air pressure inside and outside the house.

This water penetration can be reduced by providing a compartmentalized vented air space behind the cladding. The venting will keep the air pressure immediately behind the cladding equal to the outside air pressure, so that water will not be driven into the wall. The air space also creates a break in the pore structure of the cladding material, reducing water penetration by capillary action.

In this way, the cladding itself becomes a "rainscreen" (see Figure 8.4). The rainscreen will keep water from penetrating into the assembly to damage the structural components or to reduce the performance of insulation. Blocking the air space at suitable intervals (compartmentalization) will reduce lateral air flow, making the system even more effective.

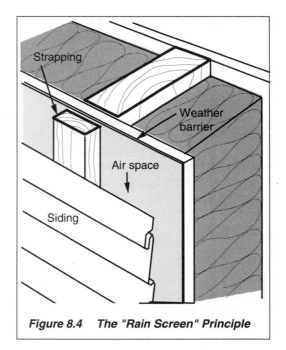

Figure 8.4   The "Rain Screen" Principle

Wind-driven penetration through cracks and holes in the cladding can be minimized by using battens, splines, or overlapping joints where cladding materials are joined.

Rainscreens and wind control are only part of good moisture control in walls. Ensuring that your wall systems perform as expected and are not adversely affected by premature deterioration requires design in accordance with sound building science principles and build then good construction practices.

- Use the driest framing material available.

- Minimize moisture penetration from the interior and exterior by making sure vapour diffusion retarders, air barriers and weather barriers are installed properly.

- Allow the system to dry to the exterior by ensuring that it is more permeable to water vapour flow on the outside than on the inside.

8.4 SINGLE STUD WALLS WITH EXTERIOR INSULATING SHEATHING

8.4.1 Description

A common way to thermally upgrade single stud walls is to add insulating sheathing (rigid or semi-rigid boardstock insulation) on the outside. Some systems also use a modified balloon framing technique to reduce thermal bridging. Exterior sheathing can be used with either a membrane or a rigid air barrier approach.

A simple 38x140 mm (2x6 in.) single stud wall using exterior insulating sheathing is shown in Figure 8.5. This type of wall system requires only minor changes to standard construction practices, and the type and thickness of sheathing can be varied without major changes to the process.

- Place 38x140 mm (2x6 in.) studs 600 mm (24 in.) on centre, depending on the requirements of the siding.

- Replace structural sheathing with diagonal bracing (wood or preformed metal) and insulating sheathing.

- Recess header joists to allow higher insulation levels at the joist assembly.

- Install a vapour diffusion retarder throughout the building envelope.

- Install a continuous air barrier to eliminate uncontrolled air exchange between indoors and outdoors.

8.4.2 Advantages and Disadvantages

This technique has several advantages.

- It uses readily available construction materials.

- Incremental costs over standard framing are relatively small.

However, it also has limitations.

- This method cannot practically provide a wall insulation value higher than RSI 4.9 (R-28).

- All electrical boxes and plumbing penetrations require separate sealing.

- Some exterior insulated sheathings can cause backing problems for the attachment of siding.

- There is a requirement to build out around window and door openings.

8.4.3 Further Considerations

Exterior Sheathing Options

A variety of boardstock insulations are available (see Chapter 5). In selecting products, consider:

- insulation levels required
- means of mechanical attachment
- product durability during construction
- cost of the installed system
- air and vapour permeability of the material and its sealing requirements
- compatibility of the sheathing with the proposed siding installation, and
- product availability.

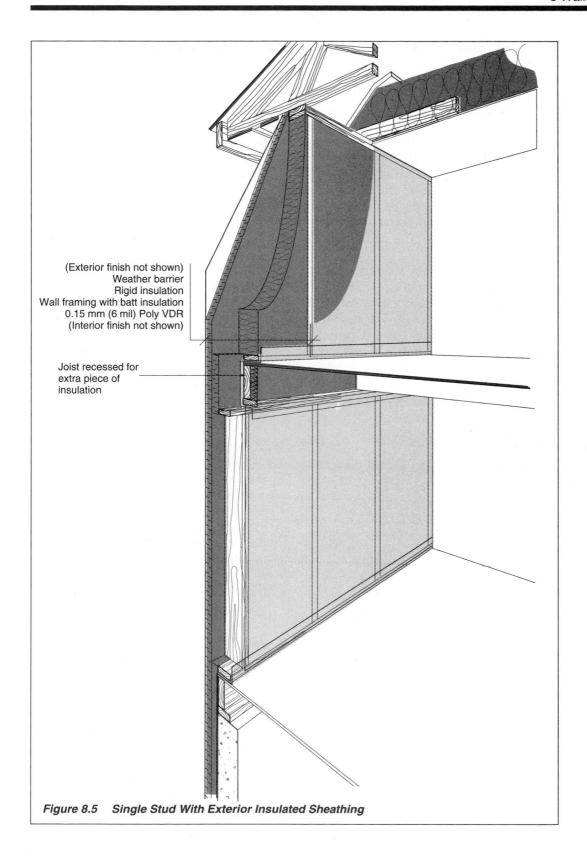

(Exterior finish not shown)
Weather barrier
Rigid insulation
Wall framing with batt insulation
0.15 mm (6 mil) Poly VDR
(Interior finish not shown)

Joist recessed for
extra piece of
insulation

Figure 8.5   Single Stud With Exterior Insulated Sheathing

Upgraded Insulation Between the Studs

The space between the studs is usually filled with batt type insulation. However, spray type materials are also suitable: these materials provide a uniform insulation level and help to reduce air leakage. Some spray applied materials may also provide higher thermal resistance per unit thickness than the batt type insulations. Since they are more dense than batt type insulations, they tend to reduce structure-borne sound transmission through the wall assembly as well.

Spray insulations require the expertise of a trained and knowledgeable applicator. In some cellulose applications, a mesh retainer is installed to hold the insulation in place. Spray foams may require trimming to ensure a flat surface after the materials have expanded in place.

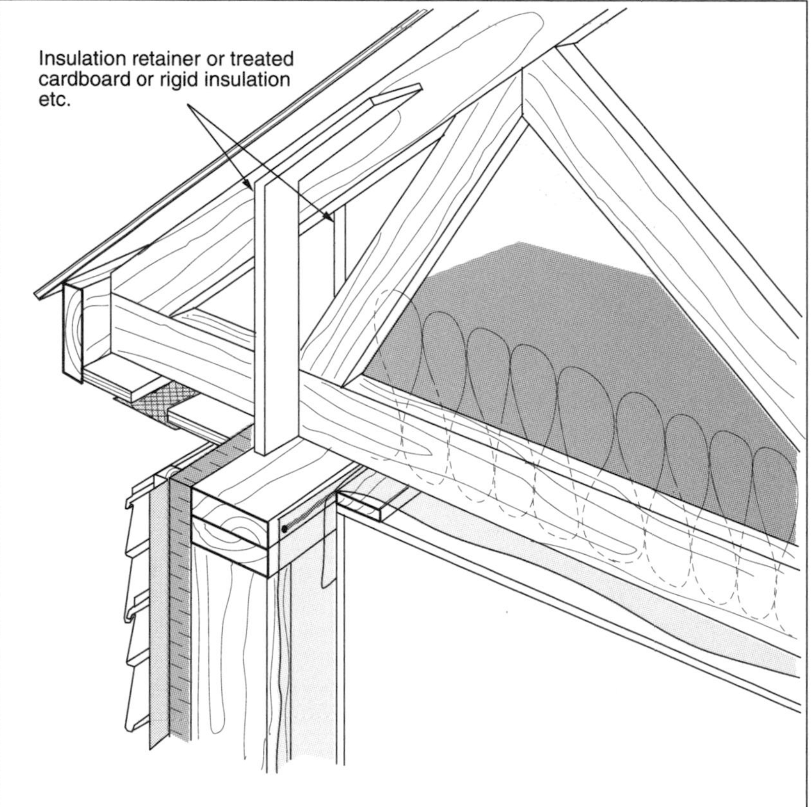

Insulation retainer or treated cardboard or rigid insulation etc.

Figure 8.6   Wall/Ceiling Junction: Single Stud Exterior Insulated Wall

The installer will assume responsibility for the appropriate density and coverage. The builder may need to leave the walls exposed for a longer period than normal to allow for adequate drying or curing.

8.4.4   System Considerations

General Air Barrier Considerations

- The air barrier must be made continuous at window and door openings, electrical boxes, plumbing and all other penetrations.

- The air barrier must also be continuous at foundations and intermediate floors. If using polyethylene or another flexible sheet material, extend it continuously around floor joist assemblies and protect it from damage during construction. (Remember: no more than one third of the RSI value of the wall should be on the warm side of the polyethylene.)

Wall/Ceiling Junction

- Ensure that the air barrier is continuous at wall/ceiling junctions. If using polyethylene, seal the ceiling polyethylene to the wall air barrier before installing the drywall (see Figure 8.6). Before installing partition walls, consider cross strapping ceiling joists to provide a space for electrical wiring and allow for a continuous polyethylene barrier.

- To restrict air flow into roof insulation, extend exterior sheathing above the soffit or install an insulation retainer.

Windows and Doors

Procedures for installing windows and doors in this type of wall system are discussed in Chapter 11. However, framing requirements must be taken into consideration prior to the installation of the exterior insulating sheathing.

Partition Walls

There are several approaches to installing partition walls.

- Frame the house with horizontal blocking installed in the outside wall between the studs where partition walls intersect. Wrap a strip of polyethylene around the last stud of the partition wall before the wall is nailed to the blocking (see Figure 8.7), and install drywall clips to support the drywall.

- Recess the partition wall from the exterior wall by at least 16 mm (5/8 in.). Install polyethylene continuously over the exterior wall and pass the drywall through the space between the partition and the exterior wall (see Figure 8.8). This eliminates the need for drywall clips. Alternatively, ensure that the drywall is continuously caulked or gasketed on both sides of the interior partition.

- Frame and insulate the exterior walls, install the polyethylene and/or hang the drywall (see Figure 8.9). Then erect all interior partitions. This approach requires clear span floors and ceilings but will eliminate air leakage from partition walls into the attic or roof space.

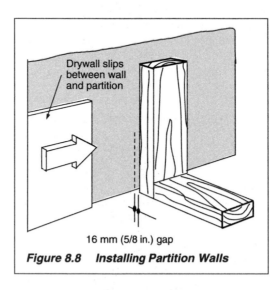

Figure 8.8   Installing Partition Walls

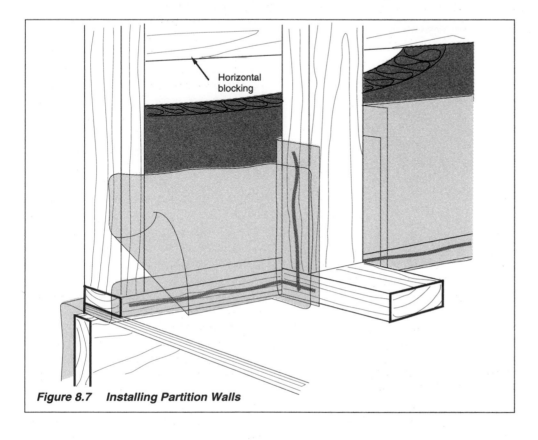

Figure 8.7   Installing Partition Walls

- Frame and insulate the exterior walls; frame the interior walls with a single top plate. Install the ceiling polyethylene in one sheet, resting on top of the interior partitions, and secure it to the roof truss with a 19x89 mm (1x4 in.) strapping. Insert a 19x89 mm (1x4 in.) strapping between the 19 x 89 mm (1x4 in.) ceiling strapping and the top of the interior partition walls.

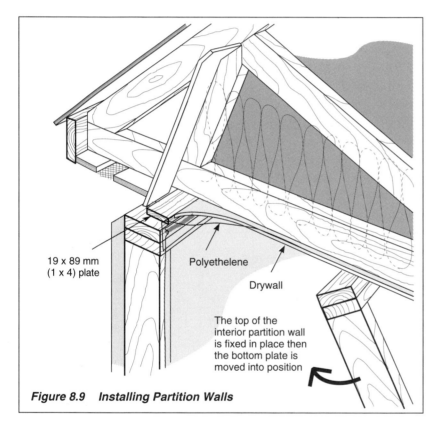

19 x 89 mm (1 x 4) plate

Polyethelene

Drywall

The top of the interior partition wall is fixed in place then the bottom plate is moved into position

Figure 8.9   Installing Partition Walls

8.5   SINGLE STUD WALLS WITH INTERIOR STRAPPING

8.5.1   Description

In this method, 38x38 mm (2x2 in.) or 38x64 mm (2x3 in.) strapping is installed vertically or horizontally on the interior of stud walls. This approach can be used with either rigid or membrane air barriers. If polyethylene is used, it is usually placed on the inside face of the studs, immediately behind the strapping.

8.5.2   Advantages and Disadvantages

This technique has several major advantages for both builders and trades.

- The combination of insulation in a 38x140 mm (2x6 in.) main wall and a 38x64 mm (2x3 in.) strapped space provides a thermal resistance of RSI 4.9 (R-28).

- The extra layer of insulation covers the inside face of the studs and reduces heat transmission.

- Electrical services can be installed on the interior side of the strapped space to avoid penetrations.

- Studding and strapping spacing can be adjusted to suit the nailing requirements of both interior and exterior finishes.

However, there are also disadvantages to this method.

- If horizontal interior strapping is used, blocking must be located so as to provide support for kitchen cabinets, baseboards, curtain rods and other fixtures.

- If vertical interior strapping is used, it will have to be drilled to allow the passage of wiring, unless it is staggered with the wall studs.

8.5.3 Possible Variations

Single stud walls with interior strapping can also be built with insulating sheathing on the exterior (see Section 8.4). This could bring the total RSI value of the wall to between 6.1 and 7.0 (R-35 and R-40), depending on materials used.

8.5.4 System Considerations

Strapping and Insulation

In addition to the normal centres, strapping is required:

- as a backer for drywall around windows and doors
- as a backer for drywall at inside and outside corners, or use drywall clips to support the drywall

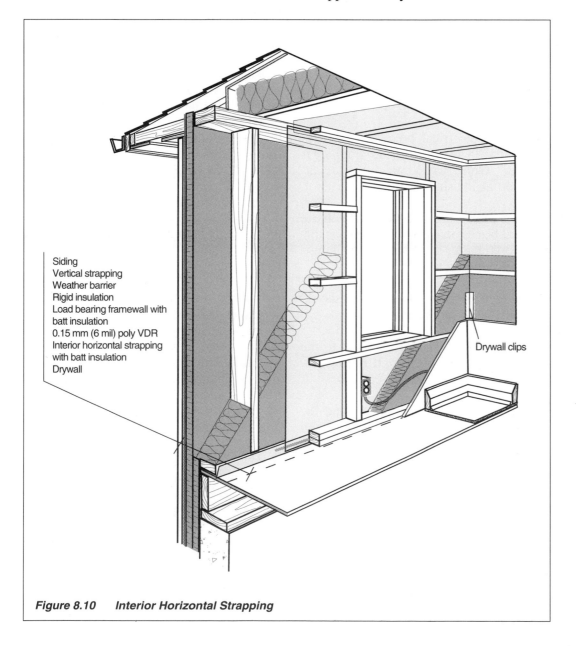

Siding
Vertical strapping
Weather barrier
Rigid insulation
Load bearing framewall with batt insulation
0.15 mm (6 mil) poly VDR
Interior horizontal strapping with batt insulation
Drywall

Drywall clips

Figure 8.10   Interior Horizontal Strapping

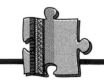
- as a support for kitchen cupboards, closet rods, etc.
- 89 mm (3-1/2 in.) above the flooring if baseboards are used, and
- as additional blocking for electrical boxes.

RSI 1.4 (R-8) batts can be friction fit into the cavities created by 38x64 mm (2x3 in.) strapping. Standard batt sizes are not generally available for 38x38 mm (2x2 in.) strapping, although some commercial products are available by special order.

Electricals and Plumbing

Electrical work must be modified to fit in the strapping space. In many cases the contractor's work will be easier.

- Shallow electrical boxes will be required if 38 mm (2 in.) wide strapping is used. Only two connections are allowed inside shallow boxes.

- Plastic boxes are preferable. If you use metal boxes, make sure the vapour diffusion retarder is not broken by clamps and grounding screws.

- Wiring runs from floor-to-floor should be drilled through the strapping.

- Wiring staples should be fixed to the cross strapping not through the polyethylene.

- A gap between the strapping and the wall top plate makes wiring runs from wall to ceiling easier, by reducing the need for drilling.

If a polyethylene air barrier is used, it must be protected from soldering carried out during plumbing work.

8.6   DOUBLE STUD WALLS

8.6.1   Description

These expanded wall systems were developed to accommodate very high levels of insulation, to ensure that polyethylene vapour diffusion retarders would not be penetrated by electrical or plumbing systems, and to provide for a continuous air seal throughout the building envelope. Expanded wall systems should be considered if the climate is extremely cold, fuel costs are exceptionally high, or the cost of insulating sheathings is very high.

Expanded wall systems consist of a load-bearing, structural wall and a lighter, non-bearing wall section that supports either exterior siding or interior drywall. The thickness of the wall assembly, and therefore the amount of insulation that can be installed, is up to the builder. This approach can be used with either membrane or rigid air barriers.

8.6.2   Advantages and Disadvantages

Expanded walls have several advantages.

- Conventional platform framing techniques can be used to construct the load-bearing wall.

- Smaller dimensioned lumber can be used.

- Flexible spacing can be used to accommodate the requirements of the cladding system.

- The method allows for a wide range of insulation values.

- Thermal bridging through the wood frame is minimized.

- Penetrations are minimized, and good air barrier continuity can be maintained.

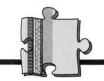

However, expanded wall systems have their disadvantages.

- Labour and material costs will be higher than for other systems.

- Platform framing techniques may not be suitable for erecting the exterior wall.

8.6.3 Possible Variations

Interior Load-bearing

One double stud wall system incorporates an interior load-bearing wall and an exterior non-bearing wall which supports exterior finishes (see Figure 8.11). The two stud walls are connected by wood panel sheathing top and bottom plates and are spaced to accommodate the required thickness of insulation. If polyethylene is used, it can be installed on the outside of the inner stud wall and protected by solid panel type sheathing; this placement reduces the likelihood that it will be damaged during the construction process.

This type of wall does have disadvantages: it is costly, and the installation of insulation is weather-dependent.

Exterior Load-bearing

Alternatively, double stud walls can have an exterior structural wall, with the second wall erected independently on the house interior (see Figure 8.12). This system allows for the same high levels of insulation and eliminates major alterations to framing.

With this system, the exterior wall is framed, sheathed, and insulated. A second stud wall is framed and installed on the house interior, spaced out to create a cavity for additional insulation. If polyethylene is used as the air barrier, it may be installed on either side of the interior frame wall and supported by a solid panel type sheathing.

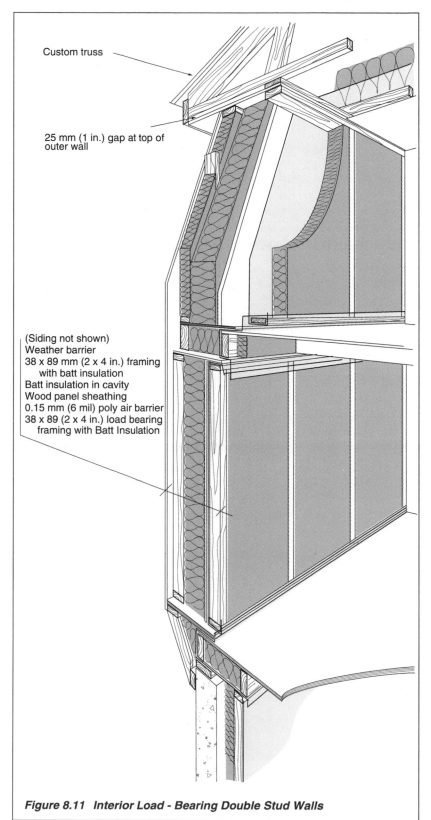

Figure 8.11 Interior Load - Bearing Double Stud Walls

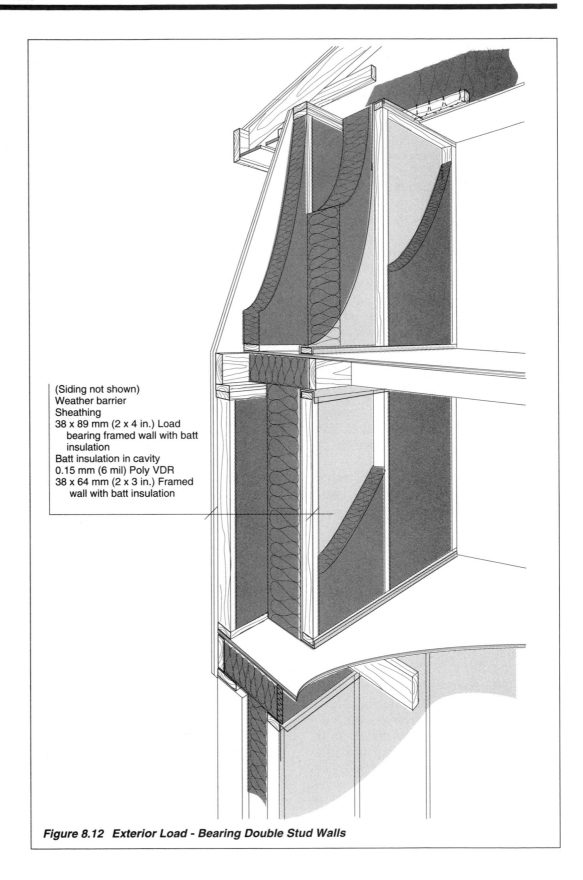

(Siding not shown)
Weather barrier
Sheathing
38 x 89 mm (2 x 4 in.) Load
   bearing framed wall with batt
   insulation
Batt insulation in cavity
0.15 mm (6 mil) Poly VDR
38 x 64 mm (2 x 3 in.) Framed
   wall with batt insulation

Figure 8.12   Exterior Load - Bearing Double Stud Walls

This system requires additional work in sealing floor joist assemblies but eliminates complex details. The interior wall, which is not load-bearing, can be framed with smaller dimension lumber such as 38x64 mm (2x3 in.) at 0.6 m (24 in.) on centre.

8.6.4   System Considerations

Windows and Doors

Expanded wall systems, as the name implies, are quite thick. Pay special attention to the location of the windows in these walls.

- If the windows are installed on the interior face of the wall, provide sloped metal flashing or an adequate slope for wood sills to promote rapid runoff.

- If the windows are installed on the outside face of the wall, detailing of sills and trim is more conventional (see Figure 8.13).

Follow the techniques described in Section 8.7 to ensure continuity of the air barrier around the windows.

Structural Considerations

Differential expansion will occur between the inner and outer wall assemblies because of the different temperatures and relative humidities to which each is subjected. Consequently, the non-bearing wall must be able to "float" relative to the load-bearing wall.

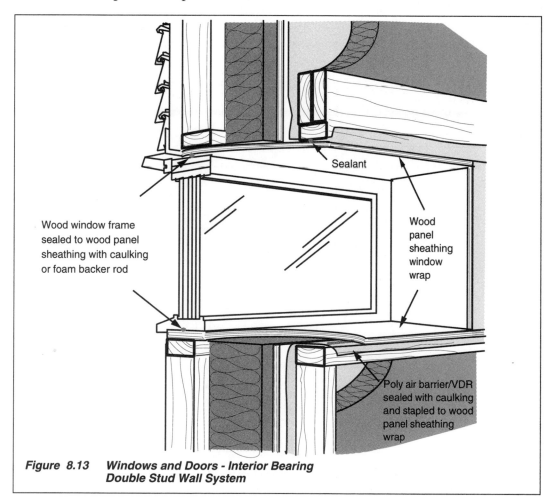

Wood window frame sealed to wood panel sheathing with caulking or foam backer rod

Sealant

Wood panel sheathing window wrap

Poly air barrier/VDR sealed with caulking and stapled to wood panel sheathing wrap

Figure 8.13   *Windows and Doors - Interior Bearing Double Stud Wall System*

For interior load-bearing double stud walls, place a 25 mm (1 in.) shim over the top plate of the interior wall to raise the truss so it is free of the outer wall.

The rough openings in the inner and outer walls are joined with a wood panel sheathing liner which provides support for interior finishes and ties the two walls together.

Insulation Options

Insulation can be installed when the wall is laid out on the subfloor as long as an adequate weather barrier is installed before the wall is erected. The insulation must fill all wall cavities. The centre cavity will generally be 89 or 140 mm (3-1/2 or 5-1/2 in.), which will accommodate RSI 2.1 or 3.5 (R-12 or R-20). This layer of insulation should be installed horizontally in the cavity and vertically between the studs.

Interior Insulated Foundation

An interior insulated foundation (shown in Figure 8.14) may be used in conjunction with double stud construction. The exterior wall assembly is supported by diagonal bracing lagged to the foundation wall. If polyethylene is being used as the air barrier, a strip of polyethylene must be installed before the floor assembly is constructed.

Inside and Outside Corner Details

Figures 8.15a and 8.15b show the inside and outside corners of an interior load-bearing double stud wall. Where partition walls intersect exterior walls, use horizontal blocking between the studs. This blocking allows insulation to run past the end of the partition.

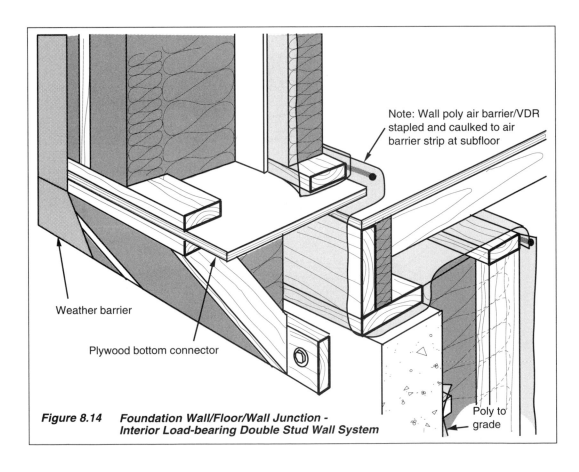

Note: Wall poly air barrier/VDR stapled and caulked to air barrier strip at subfloor

Weather barrier

Plywood bottom connector

Poly to grade

Figure 8.14   Foundation Wall/Floor/Wall Junction - Interior Load-bearing Double Stud Wall System

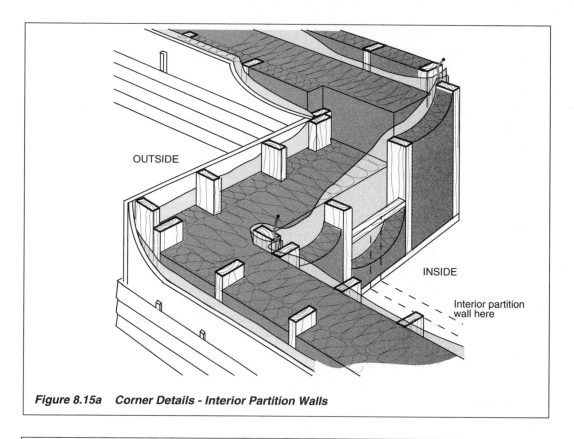

Figure 8.15a   Corner Details - Interior Partition Walls

OUTSIDE

INSIDE

Interior partition wall here

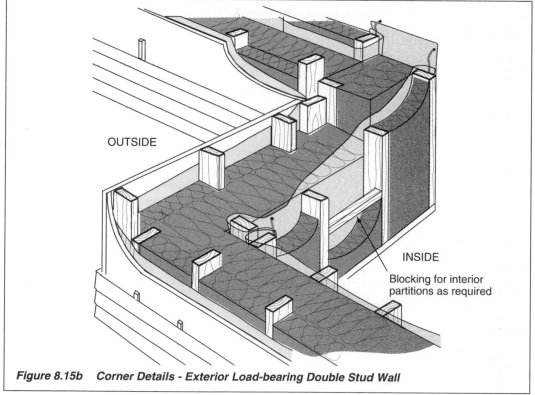

Figure 8.15b   Corner Details - Exterior Load-bearing Double Stud Wall

OUTSIDE

INSIDE

Blocking for interior partitions as required

8.7   STANDOFF WALLS

8.7.1   Description

This system uses standard framing techniques for a single stud, load-bearing wall with sheathing. Intermediate floor assemblies and partition walls are all independent.

Either a rigid or a membrane air barrier can be used with this approach. If a polyethylene air barrier is used, it should be installed over the sheathing, from the top of the completed wall to the foundation wall. A truss wall is then anchored, through the polyethylene and sheathing, into the wall studs (see Figure 8.16). Alternatively, the wood panel sheathing can be used as the air barrier as long as all joints are sealed.

Trusses can be either site-fabricated or manufactured using a variety of designs and mechanical fastening techniques. The outer member can be built of smaller dimension lumber since it is not load-bearing. Trusses should be designed to minimize gaps in the insulation behind framing members. The trusses should be spaced appropriately to accommodate the required thickness and width of insulation.

8.7.2   Advantages and Disadvantages

This method has several advantages.

- The load-bearing frame uses standard framing.

- When horizontally strapped, the wall can be designed to accommodate high insulation levels and can be used with vertical siding.

- A polyethylene air barrier can be applied as one large single sheet, with a minimum number of seams and penetrations.

However, like other methods, it also has disadvantages.

- Gaps may occur in the insulation if trusses are not properly spaced and batt insulation is not properly installed.

- On buildings of more than a single storey, scaffold work is required.

- Differential shrinkage between the inner and outer truss members and the header joist can cause deformation and cracking of the interior finish material at the ceiling/wall joint.

- The vapour diffusion retarder and insulation materials are exposed to the weather during installation.

8.7.3   System Considerations

Materials Considerations

If using polyethylene, use an ultraviolet-stabilized material as it may be exposed to sunlight for a long period during the construction process.

Rigid board insulation can be installed between chords of the standoffs to minimize air pockets in the insulation.

Structural Considerations

If the wall trusses are nailed to the rafter or truss tails, the base should be free floating to accommodate differential movement.

Windows and Doors

Figure 8.17 shows the detail of a wood panel sheathing window box in a standoff wall system.

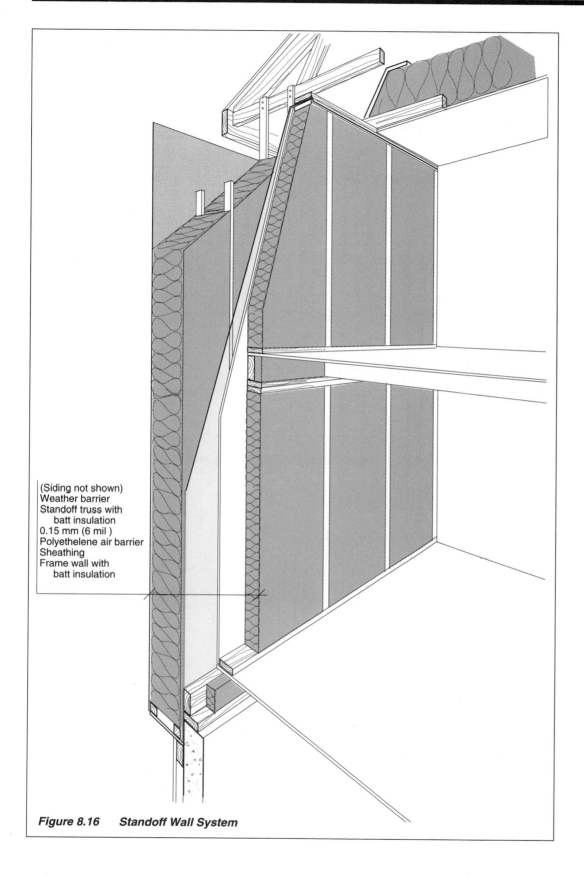

(Siding not shown)
Weather barrier
Standoff truss with
     batt insulation
0.15 mm (6 mil)
Polyethelene air barrier
Sheathing
Frame wall with
     batt insulation

Figure 8.16   Standoff Wall System

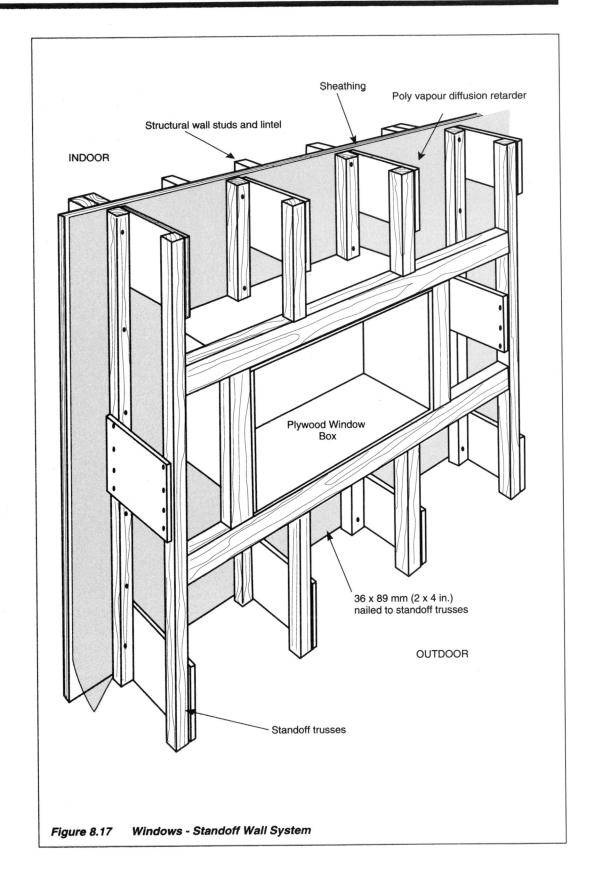

Figure 8.17   Windows - Standoff Wall System

8.8 RIGID INSULATING CORE PANEL WALLS

8.8.1 Description

These systems use foam insulation panels that have locking details at the edges so that joints fit together to make an airtight seal. Standard framing techniques can then be used to erect a load-bearing wall system.

Manufacturers use various techniques to build the framing into the wall system while providing for electrical wiring and other services. One method is illustrated in Figure 8.18. The studs are set into pregrooved channels that provide a 50 mm (2 in.) gap in front of the insulation to run wires while still providing a complete thermal break behind the studs. If greater insulating value is desired, this space can be filled with insulation prior to installing the interior finish.

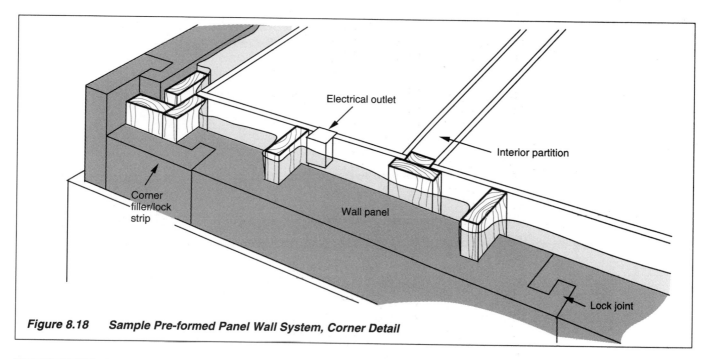

Electrical outlet

Interior partition

Corner filler/lock strip

Wall panel

Lock joint

Figure 8.18   Sample Pre-formed Panel Wall System, Corner Detail

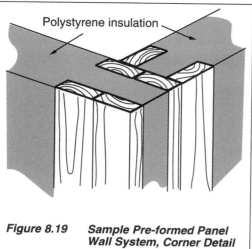

Polystyrene insulation

**Figure 8.19   Sample Pre-formed Panel
Wall System, Corner Detail**

Figure 8.19 shows another system. Note that insulation is placed between the framing members to provide a thermal break. This system must be cut or grooved to make room for electricals and plumbing. Doors and windows can be cut into the system on site, and framed using standard methods.

8.8.2   Advantages and Disadvantages

Rigid insulating core panel walls have several advantages.

- Uniform insulation of the wall assembly is assured; there are no voids.

- A good air barrier is provided if the joint system seals well.

- A thermal break is provided behind all studs.

- Installation labour requirements are lower since the studs and insulation are erected at the same time.

However, the panels also have disadvantages.

- Other trades may have to cut or groove the material to install wiring, plumbing, etc.

- Plumbers and other trades must be careful not to ignite the insulation.

- Thermal bridging may occur through top and bottom plates.

- Drawings and estimates must be accurate to ensure accurate assembly on site.

8.9   STRESS-SKIN PANELS

8.9.1   Description

Stress-skin panels are glass fibre, polystyrene or polyurethane insulating cores sandwiched between skins of wood, wood panel sheathing, waferboard, or drywall (see Figure 8.20). The panels come in a wide range of configurations depending on the insulating core, the type of skin applied to either side, and whether or not "stiffeners" are installed in the panels to make them more rigid. Some systems can be installed over structural timber frames, some can be used as sheathing over wall or roof systems, and some can act as the entire shell of the house, including the structural element.

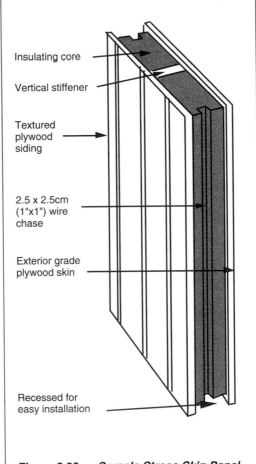

Insulating core

Vertical stiffener

Textured plywood siding

2.5 x 2.5cm (1"x1") wire chase

Exterior grade plywood skin

Recessed for easy installation

Figure 8.20   Sample Stress Skin Panel

8.9.2 Advantages and Disadvantages

Stress-skin panels have a number of advantages.

- Uniform insulation of the wall assembly is assured; there are no voids.
- This method provides a good air barrier if the joint system seals well.
- The panels provide structural support to the wall system.

However, they do have disadvantages as well.

- Other trades may have to cut or groove the material to install wiring, plumbing, etc.
- Plumbers and other trades must be careful not to ignite the insulation.
- There may be thermal bridging through top and bottom plates, and through vertical "stiffeners" if they are used.
- Accurate drawings and estimating are essential to ensure accurate assembly on site.

8.9.3 Possible Variations

Most of the panels use polystyrene as the insulating core. It is cheaper, but offers a lower RSI (R) value per unit thickness than does polyurethane. Polystyrene and polyurethane cores also vary in terms of fire resistance and durability.

- Polystyrene melts at about 93°C (200°F) and flows readily at 121°C (250°F), so a polystyrene system will fail in a fire even if the panels don't burn. Polyurethane, on the other hand, will stay intact until it burns. Either system can be made to comply with the requirements of the fire code: some manufacturers install firestops throughout panels made with polystyrene cores to improve performance.
- Low-density polystyrene is soft and breakable, so the adhesive used to fasten it to the skins must penetrate the insulation to ensure a good bond.

The surface of polyurethane insulation does not readily promote bonding because the cells of the insulation are cut open, making it difficult to achieve even adhesion unless the insulation is foamed in place. In both cases, polyurethane adhesives generally perform best.

Most stress-skin systems use a wood spline to join panels together (see Figure 8.21). However, full splines act as thermal bridges and can cause cracking on interior finishes if their dimensions change due to changes in moisture content. Some systems use a double-spline system; the splines are installed with their long dimension parallel to the wall surface, leaving room for insulation between them. The effectiveness of this system depends on the thickness of the panel (the thickness of the insulation between the splines) and the seal created by the joint.

8.9.4 System Considerations

Panels may require strapping to accommodate heavy siding materials because there are no studs. As well, stress-skin panel systems must make provision for electrical wiring and other services. Better-resolved systems provide raceways, i.e. grooves precut into the core, for wire.

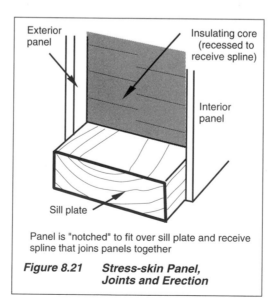

Panel is "notched" to fit over sill plate and receive spline that joins panels together

Figure 8.21   ***Stress-skin Panel, Joints and Erection***

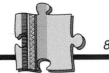

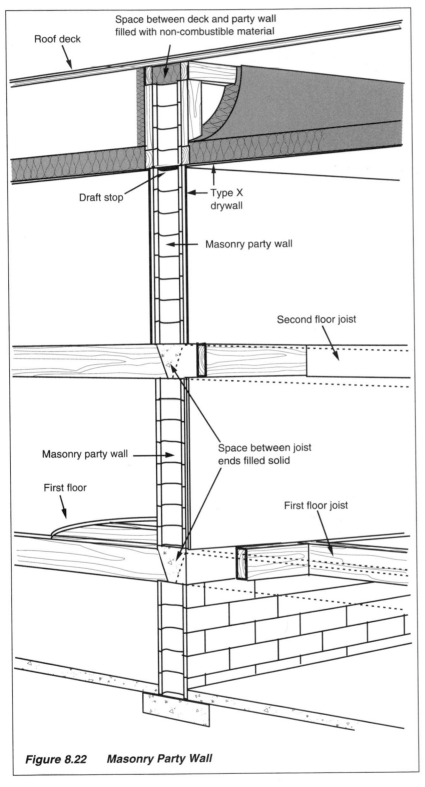

Roof deck

Space between deck and party wall filled with non-combustible material

Draft stop

Type X drywall

Masonry party wall

Second floor joist

Masonry party wall

Space between joist ends filled solid

First floor

First floor joist

Figure 8.22   Masonry Party Wall

8.10   PARTY WALLS IN MULTI-UNIT BUILDINGS

Party walls can be built with either non-combustible materials or a combustible assembly with a fire rating of one hour. They may terminate at the underside of the roof sheathing. Figure 8.22 and 8.23 illustrate the two methods most commonly used to build party walls.

This section focuses on the techniques which can be used to make these walls reasonably soundproof and smoke-tight while reducing heat loss.

Air barriers are installed to provide a continuous barrier for each unit, as though it were not attached to any other.

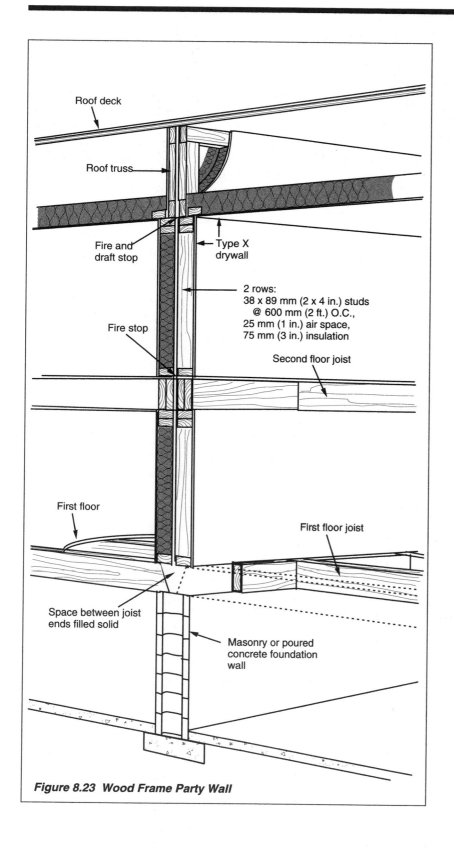

Figure 8.23 Wood Frame Party Wall

Labels on figure:
- Roof deck
- Roof truss
- Fire and draft stop
- Type X drywall
- Fire stop
- 2 rows: 38 x 89 mm (2 x 4 in.) studs @ 600 mm (2 ft.) O.C., 25 mm (1 in.) air space, 75 mm (3 in.) insulation
- Second floor joist
- First floor
- First floor joist
- Space between joist ends filled solid
- Masonry or poured concrete foundation wall

8.10.1 Exterior Wall Junctions

Block Walls

Anchor rods should be installed every 0.9 m (36 in.) in a block wall while it is being erected to tie the exterior wall to the party wall. Figure 8.24 shows how the joint between the party wall and the exterior wall can be sealed using either the mem-
brane (right side of the illustration) or the rigid (left side) air barrier approach. The end of the masonry wall should be covered with non-combustible insulation to prevent it from acting as a thermal bridge. Firestops such as 12.7 mm (1/2 in.) drywall or equivalent must be installed on *both* sides of the truss that separates two adjacent attic spaces. These firestops must extend out to the eaves.

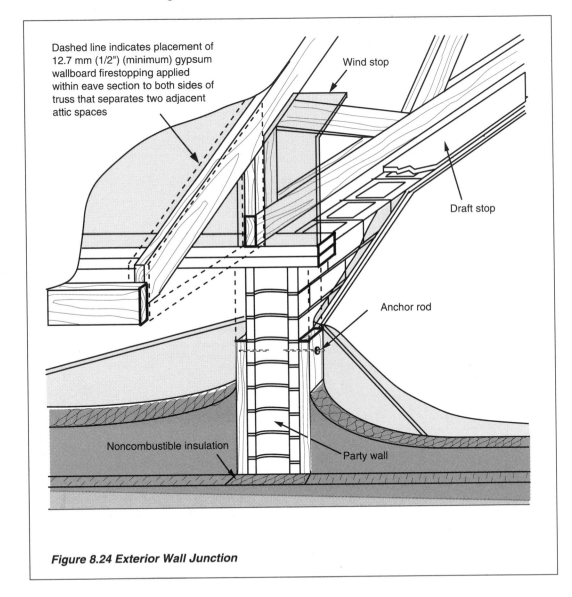

Dashed line indicates placement of 12.7 mm (1/2") (minimum) gypsum wallboard firestopping applied within eave section to both sides of truss that separates two adjacent attic spaces

Wind stop

Draft stop

Anchor rod

Noncombustible insulation

Party wall

Figure 8.24 Exterior Wall Junction

Where adjacent units are offset, the party wall need not extend beyond the joint with the exterior wall of the innermost unit (see Figure 8.25). A noncombustible thermal break should be installed across the end of the party wall.

Double Stud Walls

The outermost stud spaces of both halves of a double stud party wall should be insulated in order to ensure continuity of the exterior insulation across the party wall. The 25 mm (1 in.) air space between the two rows of studs should be sealed (see Figure 8.26). This can be done with a metal firestop, or by using firestop material as the sheathing.

Where adjacent units are offset, the double stud party wall need not extend beyond the joint with the exterior wall of the inner-most unit (see Figure 8.27). A firestop should be installed across the 25 mm (1 in.) air space between the rows of studs. Both rows of studs should be insulated at least one stud space into the party wall. If combustible siding is used over a wood stud party wall extension, a double layer of 12.7 mm (1/2 in.) drywall sheathing should be used to provide the necessary 3/4-hour fire rating. This is not needed with brick veneer.

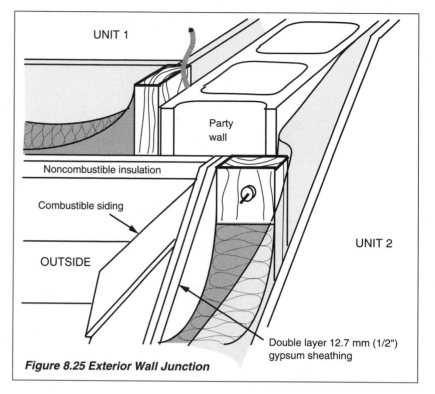

Figure 8.25 Exterior Wall Junction

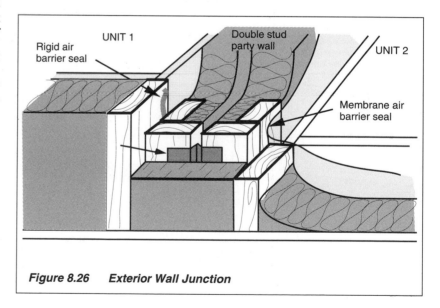

Figure 8.26   Exterior Wall Junction

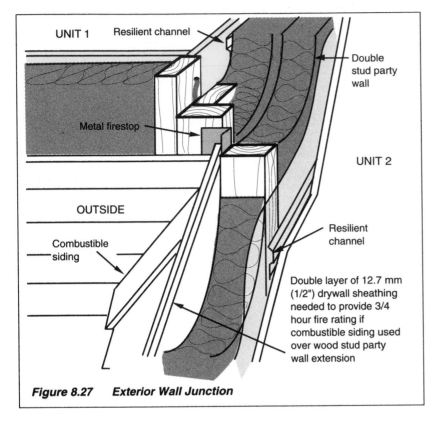

Figure 8.27   Exterior Wall Junction

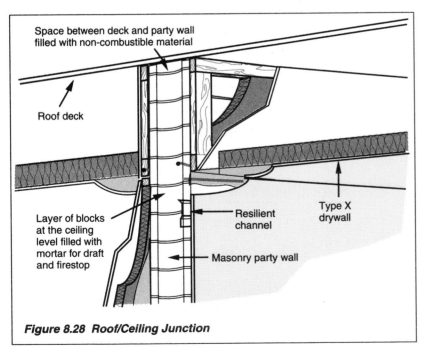

Figure 8.28 Roof/Ceiling Junction

8.10.2 Attic Penetrations

Block Walls

Figure 8.28 shows how to eliminate air leakage at the ceiling/party wall junction. The layer of blocks at the ceiling level must be filled with mortar or solid masonry to prevent moisture-laden air in the cores from entering the attic space. A row of solid blocks would also serve the purpose. The blocks should terminate just below the line of the roof deck. When the trusses and sheathing are installed, a layer of noncombustible insulation material should be installed over the tops of the blocks to provide both a firestop and a thermal break. Both faces of the party wall should be insulated to prevent thermal bridging through the masonry.

The joint between the ceiling and the wall must be sealed to stop air leakage into the attic. If a membrane air barrier is used, it should be sealed to the polyethylene on the wall surface. If drywall is being used as a rigid air barrier, the joint will be sealed when the corner is taped and finished. Note that in either case, all penetrations into the space behind the drywall and into the party wall above the ceiling must be sealed to keep air from leaking into these spaces and rising into the attic.

Double Stud Walls

Figure 8.29 shows how to eliminate air leakage into the attic space if a double stud party wall is used. A metal firestop should be used to block air leakage through the 25 mm (1 in.) air space. Fire-rated gypsum board should be used to continue the fire rating up to the underside of the roof sheathing. Insulation can be installed in the space created by the trusses above the top plate. This will reduce heat loss into the attic space.

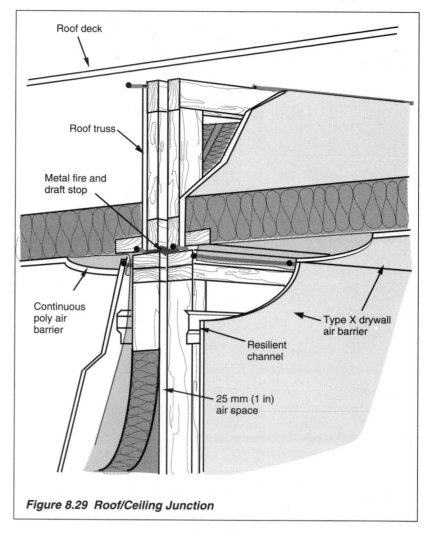

Roof deck

Roof truss

Metal fire and
draft stop

Continuous
poly air
barrier

Type X drywall
air barrier

Resilient
channel

25 mm (1 in)
air space

Figure 8.29 Roof/Ceiling Junction

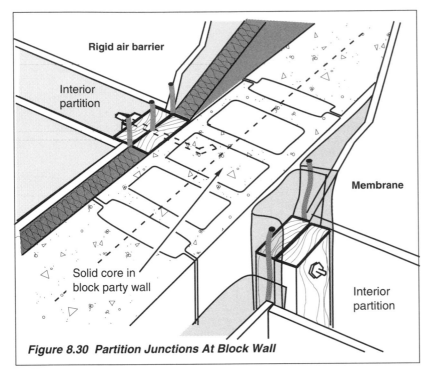

Figure 8.30 Partition Junctions At Block Wall

8.10.3 Load-bearing Partition Wall Junctions

Block Walls

Load-bearing partition walls should be anchored to masonry party walls as shown in Figure 8.30. If a membrane air barrier is used, it should be installed between the wall stud and the party wall. If a rigid air barrier is used, the edges of the drywall on both sides of the stud next to the party wall must be sealed to the stud. The joint will be made fully airtight when the drywall installed on the partition wall is taped and finished to the sheet on the party wall.

Double Stud Walls

Backing should be placed in wood stud party walls to accommodate the partition walls that are erected later in the process (see Figure 8.31). The techniques described in Section 8.4.4 can be used to make the air barrier continuous.

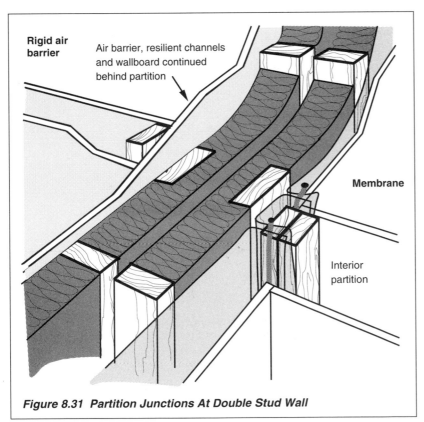

Figure 8.31 Partition Junctions At Double Stud Wall

8.11 SERVICE PENETRATIONS

This section describes several techniques which can be used to ensure the continuity of the air barrier around service penetrations in exterior walls.

8.11.1 Plumbing System Components

Two types of plumbing components may penetrate exterior walls: drain pipes and vent stacks (usually plastic), and water supply pipes (usually copper). If possible, these pipes and stacks should be kept inside the air barrier. However, if you use the rigid air barrier approach, or if plumbing penetrates a membrane air barrier, the penetrations must be sealed to the air barrier to prevent air leakage into the insulated space, allowing for the expansion and contraction characteristics of the materials.

Where plumbing components will pierce the air barrier, mount a wood panel sheathing backing board flush with the inside face of the studs (see Figure 8.32 a). Drill holes through the wood panel sheathing in the locations where the pipes will penetrate. Bear in mind that if the holes are too small, the pipes may rub against the edge, causing squeaks whenever the hot water is used. Seal the gap between the pipe and the edge of the hole with caulking or another suitable sealant material.

If using a rigid air barrier approach, mount the wood panel sheathing onto the studs so it will be flush with the face of the drywall. Horizontal blocking should be placed between the studs as a backing for the seam between the sheathing and the drywall. This seam must be sealed to ensure continuity of the air barrier (see Figure 8.32 b).

Another rigid air barrier approach is to construct an air tight box which the plumbing penetrates vertically (see Figure 8.32 c). Drywall is then sealed to the face of the box using foam tape.

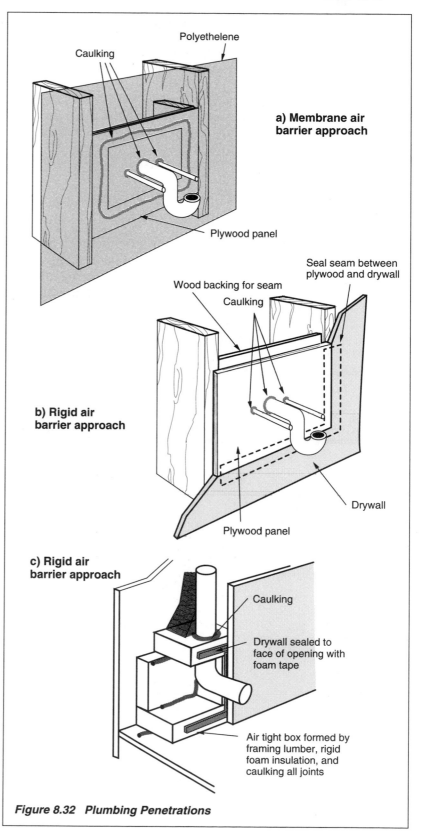

a) Membrane air barrier approach

Caulking

Polyethelene

Plywood panel

b) Rigid air barrier approach

Wood backing for seam

Caulking

Seal seam between plywood and drywall

Plywood panel

Drywall

c) Rigid air barrier approach

Caulking

Drywall sealed to face of opening with foam tape

Air tight box formed by framing lumber, rigid foam insulation, and caulking all joints

Figure 8.32 Plumbing Penetrations

8.11.2 Electrical System Components

The electrical system may penetrate the air barrier at the following locations:

- main service conduits and the main service box if it is located on an exterior wall
- wall top plates and end studs of partition walls where wires pass through
- electrical outlets and light switch boxes, and
- exterior outlets and lights.

These penetrations must be sealed in some way to maintain continuity of the air barrier. Wires and conduits running to the main service box can be sealed by caulking them to a wood panel sheathing backing board on which the main service box is mounted. Wiring passing through top plates and end studs can also be sealed with caulking. Electrical boxes can be placed in site built or prefabricated airtight boxes (see Figure 8.33).

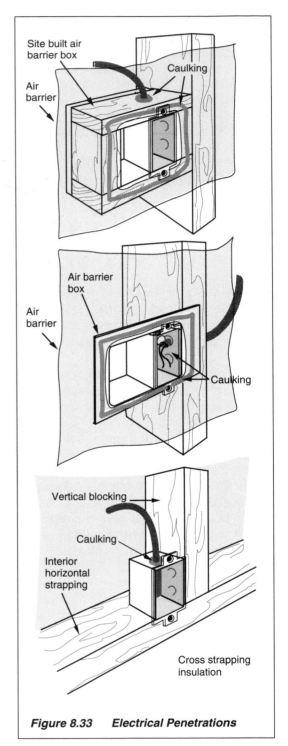

Figure 8.33   Electrical Penetrations

Wiring, electrical boxes, etc. inside the air barrier do not require separate sealing. Main service conduits passing through the air barrier from the outside need to be sealed.

Run the conduit through a wood panel sheathing plate, then seal the conduit to the wood panel sheathing and the sheathing to the air barrier (see Figure 8.34). Also seal the inside of the conduit.

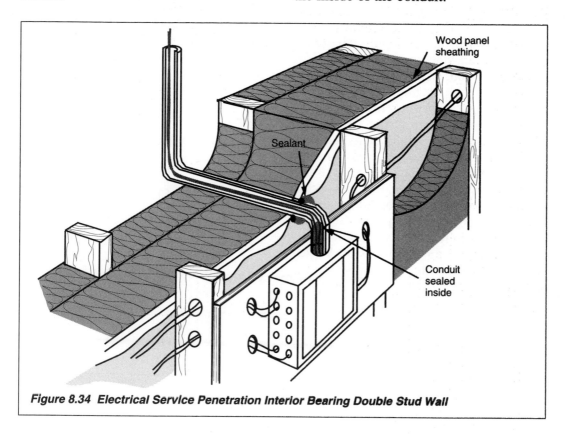

Figure 8.34 Electrical Service Penetration Interior Bearing Double Stud Wall

If you use a rigid air barrier, you should locate electrical outlets and wall switches on interior walls where possible. Those installed on exterior walls must be sealed to the gypsum board or drywall.

- Mud the gypsum board directly to the outlet and install a compressible gasket directly under the cover plate of the outlet or switch (see Figure 8.35). Use outlet boxes without openings, such as sealed plastic boxes. Wires penetrating these boxes should also be caulked to the box.

- Alternatively, to avoid penetrating the gypsum board or drywall on exterior walls, locate low-profile, surface-mounted boxes at baseboard level. Run electrical wires directly down from the outlet through the floor sheathing and into the floor joist space. Check local codes before using this approach.

8.11.3 Mechanical System Components

Ductwork for heat recovery ventilators (HRVs) and exhaust fans will also require sealing, using the techniques outlined above for electrical conduits. For warmer surfaces such as chimney flues, use sheet metal. Seal the sheet metal to the metal duct with a caulk or sealant capable of withstanding the high temperature.

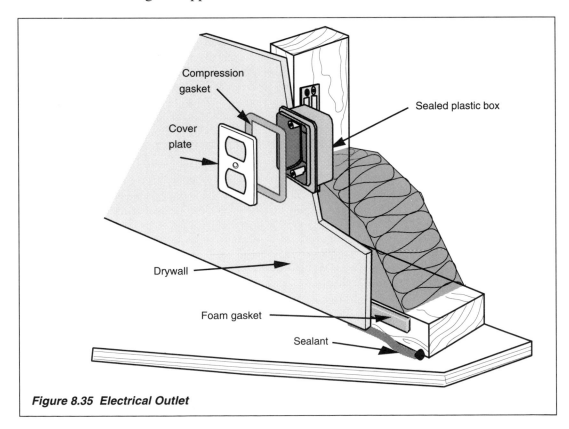

Figure 8.35 Electrical Outlet

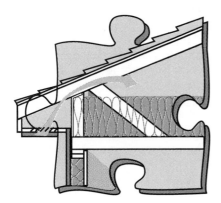

Chapter *9*

ATTICS AND ROOFS

The "roofs" of most Canadian houses actually consist of a ceiling assembly, an attic and a roof. As with other components of the building envelope, proper insulation and control of air leakage and moisture are essential.

The first section of this chapter looks at general design considerations for roofs, and ways to avoid common problems. Sections 9.2 and 9.3 consider different approaches to constructing roofs with high insulation levels. Sections 9.4 and 9.5 show how to minimize the flow of air and moisture from the house into the attic space. Finally, Section 9.6 looks at special considerations that apply in northern regions.

9.1 GENERAL DESIGN CONSIDERATIONS

9.1.1 Insulation

Roofs have traditionally been the most heavily insulated part of the house. Levels of RSI 7.0 (R-40) to RSI 10.5 (R-60) are common. Attic spaces are easy to insulate to high levels with low-cost insulations such as blown- or batt-type.

However, the quantity of insulation alone does not determine its effectiveness. The performance of roofs and attics can usually be improved by:

- installing a wind barrier at the eaves to keep wind from blowing through the insulation
- sealing joints and penetrations to eliminate or reduce air leakage around plumbing stacks, electrical fixtures, at partition wall top plates, around chimneys and flues, and around attic access hatches
- increasing insulation thickness in cathedral ceilings
- eliminating gaps in the insulation, particularly at truss struts
- maintaining full insulation thickness at the eaves, and
- reducing thermal bridges through structural members.

Proper insulation is important, but it is only part of a sound attic and roof assembly. Control of moisture-related problems is also critical to ensure good performance and durability.

9.1.2 Moisture Control

During the winter, the attic in most houses is a cold, ventilated space between the roof and ceiling insulations. In most flat roofs and cathedral roofs, the attic is replaced by a ventilation space.

In addition to actual roof leaks, moisture can enter this space in two other ways:

- Warm, moist air can leak into the attic space through air leakage paths (see Figure 9.1) and condense on cold surfaces.

- Water vapour may diffuse into the attic through ceiling building materials.

In colder regions, frost may build up in the attic space during the winter and cause serious problems in the spring when it melts.

Ventilation alone is not enough to deal with this problem. During periods of cold weather, the outside air cannot hold much water vapour. Cold ventilation air entering the space from outside cannot carry away all of the moisture that can enter the attic space.

The best way to reduce condensation in roof and attic spaces is to prevent water vapour from entering in the first place. This is done by eliminating air leakage through the ceiling and installing a vapour diffusion retarder, usually polyethylene, in the following manner:

- Carry the polyethylene across the top plates of partition walls.

- Seal around chimney and flue penetrations using fire stop spacers.

- Provide properly designed air- and va-pour-tight seals around plumbing stacks.

- Minimize electrical fixtures and penetrations in the ceiling. Use recessed lighting fixtures approved by the Canadian Standards Association (CSA) for insulated ceiling applications.

- Weatherstrip and latch attic access hatches, or place them on gable end walls outside the house.

9.1.3 Ceiling Sag

Ceiling drywall can sag because of:

- the absorption by the drywall of its water-based texture finishes
- high humidity from drying lumber and interior finishing materials
- condensation on the drywall, and
- water damage caused by wind-driven rain and snow entering the attic or roof space.

These problems can be reduced or elimi-nated by:

- using thicker drywall on ceilings
- applying oil primer before applying heavy water-based textured finishes
- providing ventilation during construc-tion
- preventing wind-driven rain and snow from entering the space, and
- insulating the attic or roof space before the heat is turned on.

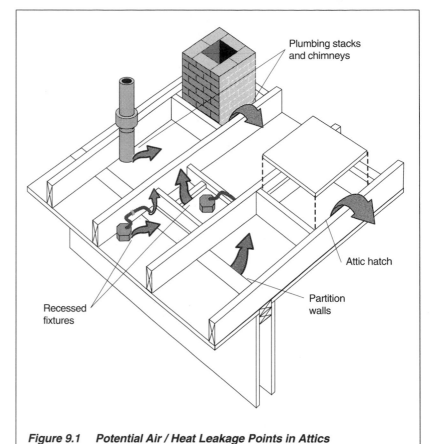

Figure 9.1   Potential Air / Heat Leakage Points in Attics

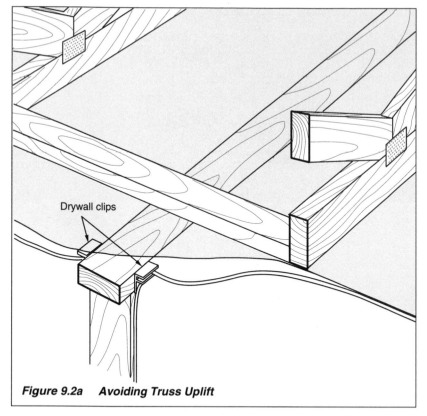

Figure 9.2a   Avoiding Truss Uplift

9.1.4   Truss Uplift

Truss uplift will occur in the winter when the upper chord of a truss contains more moisture than the lower chord. The upper chord lengthens and bows upward, pulling up the lower chord and causing the ceiling to move upward.

Truss uplift will damage interior finishes, particularly where the ceiling and interior partition walls join.

Some pieces of lumber — called "reaction wood" — expand and contract much more than the average. If the upper chord of a truss is made from reaction wood, truss uplift can occur every winter.

There are several ways to minimize the effects of truss uplift.

- Buy trusses from a manufacturer who uses properly dried lumber.

- Ensure that the attic or roof space is adequately ventilated.

- Locate attic vents to ensure good air flow.

- Ensure that soffit vents are not blocked by insulation.

- Connect the ceiling drywall to the partition walls using drywall clips and nail the ceiling drywall far enough from the partition wall intersection to allow for deflection without joint failure (see Figures 9.2a and 9.2b).

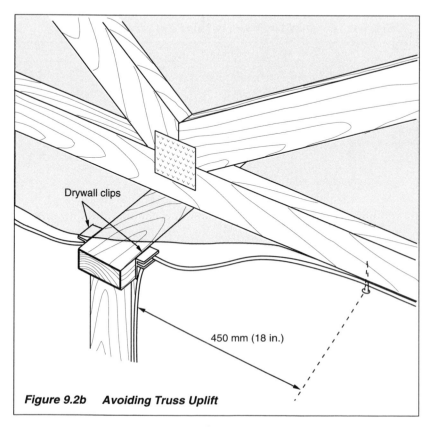

Figure 9.2b   Avoiding Truss Uplift

9.1.5 Ice Damming

Ice dams can form if the roof is covered with snow that extends to the eaves. Radiation from the sun or heat loss from the house can warm the attic or roof space, causing the snow to melt. Heat loss through the double top plate at the exterior wall/ceiling junction can be a particular problem if provision is not made for full-depth insulation.

As it runs down the roof surface, this melted snow reaches the much cooler area beyond the exterior wall, above the soffit. There, it can refreeze and cause icicles. The layer of ice at the edge of the roof thickens as this process continues, creating a "dam" that traps further melting snow. If the situation persists, water can back up under shingles and leak into the attic in the area above the top plate of the exterior wall.

Ice damming can be prevented by:

- ensuring that the entire ceiling, including the area above the top plates of the exterior walls, is protected by a full-depth layer of insulation
- eliminating heat loss caused by air leakage into the attic or roof space, and
- ventilating the attic space to prevent heat build-up under the roof sheathing.

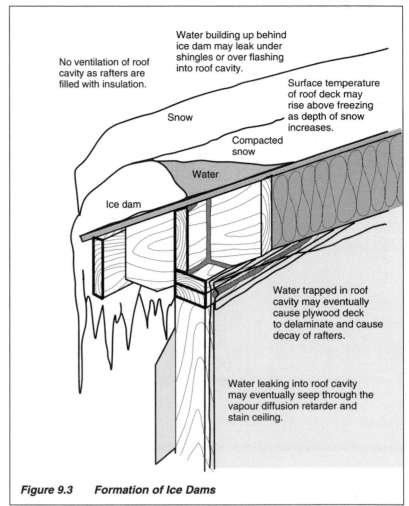

Figure 9.3   Formation of Ice Dams

9.2 TRUSSES

Most builders use prefabricated roof trusses. However, many standard prefabricated trusses severely reduce the space for insulation at the eaves. This section discusses trusses that allow higher levels of insulation to be installed in the roof.

9.2.1 Raised Heel Trusses

A raised heel truss is shown in Figure 9.4. There will be additional costs associated with this type of truss. However, energy savings and the prevention of interior surface condensation or "dusting" may justify the investment.

Advantages

- This type of truss allows for full insulation depth at the perimeter of the attic.

- A uniform thickness of insulation can be installed over the entire attic floor including above the top plates.

Disadvantages

- The initial purchase price is higher than for standard trusses.

- The design requires a higher exterior stud wall, and additional siding, which also increases costs.

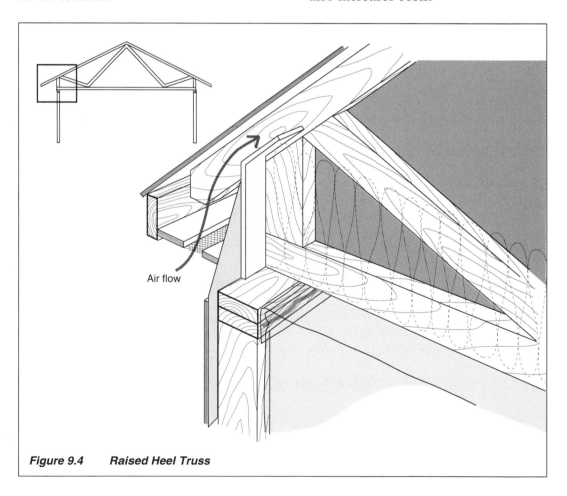

Air flow

Figure 9.4   Raised Heel Truss

9.2.2 Cantilever Truss

An alternative is the cantilevered truss, which has an additional block inserted for bearing on the top plate (see Figure 9.5). No extra siding is required. However, depending on the slope of the roof and the projected length of the cantilever, there may not be enough space above the top plate to allow for full-depth insulation.

Advantages

• A uniform thickness of insulation can usually be installed over the entire attic floor including above the top plates.

Disadvantages

• The initial purchase price is slightly higher than for standard trusses.

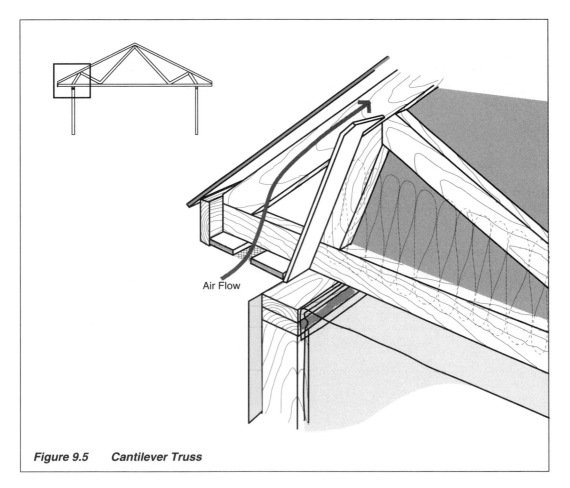

Air Flow

Figure 9.5   Cantilever Truss

9.2.3 Dropped Chord Truss

This truss has an additional bottom chord (see Figure 9.6). The space between the structural bottom chord of the truss and the dropped chord is filled with insulation. As a result, all of the structural elements of the truss are exposed to the same thermal and relative humidity conditions, i.e. outdoor conditions.

Advantages

• Dropped chord trusses can help to reduce truss uplift.

• The truss allows for full-depth insulation up to the perimeter walls.

Disadvantages

• The initial purchase price is sometimes higher than for standard trusses.

• Longer studs are required.

• There are extra costs for the additional siding required due to the higher exterior stud wall.

• Blocking is required at the ceiling/exterior wall joint.

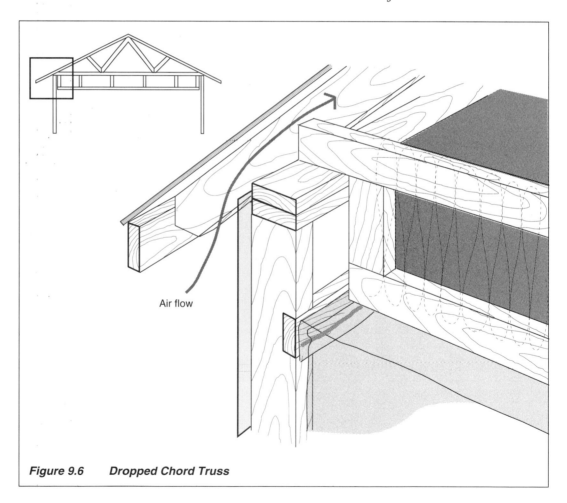

Air flow

Figure 9.6   Dropped Chord Truss

9.2.4 Scissor Truss

A scissor truss has a sloping, as opposed to horizontal, bottom chord (see Figure 9.7). These trusses provide the interior appearance of a cathedral ceiling without a central, load-bearing beam or wall.

This truss can be modified to accommodate higher levels of insulation by using a raised heel.

Advantages

- Scissor trusses are an economical way to provide vaulted ceilings.

- These trusses can more easily accommodate higher levels of insulation than conventional cathedral ceilings can.

- Full-depth insulation can be provided up to the perimeter walls.

Disadvantages

- It is difficult to insulate between the chords except with blown material.

- Baffles may be required to prevent settling of loose fill insulation.

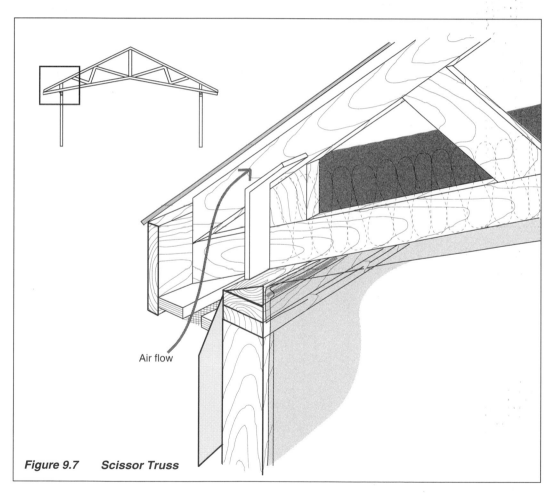

Air flow

Figure 9.7   Scissor Truss

9.2.5 Parallel Chord Truss

Parallel chord trusses consist of parallel chords of wood joined by an open web of wood or steel braces, or a solid web of wood panel sheathing (see Figures 9.8a and 9.8b). Some manufacturers install insulation between the webs and chords of this type of truss.

Advantages

- These trusses allow for the installation of large amounts of insulation in cathedral ceilings.

- Ventilation can be provided without the use of cross-purlins when open web materials are used.

Disadvantages

- If a steel web is used, heat loss due to thermal bridging can be high.

- It is difficult to insulate between chords, particularly if a solid wood web is used.

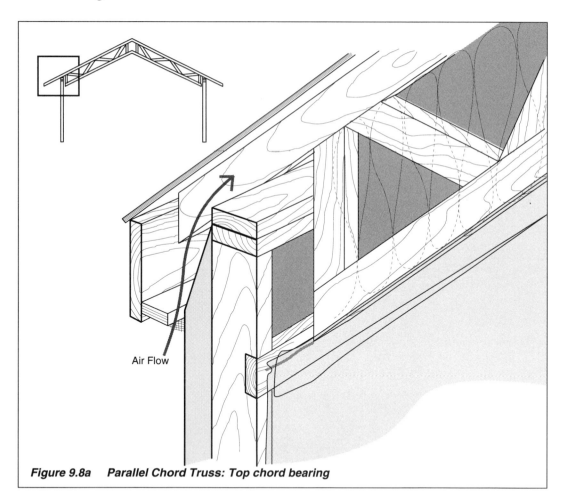

Air Flow

Figure 9.8a   Parallel Chord Truss: Top chord bearing

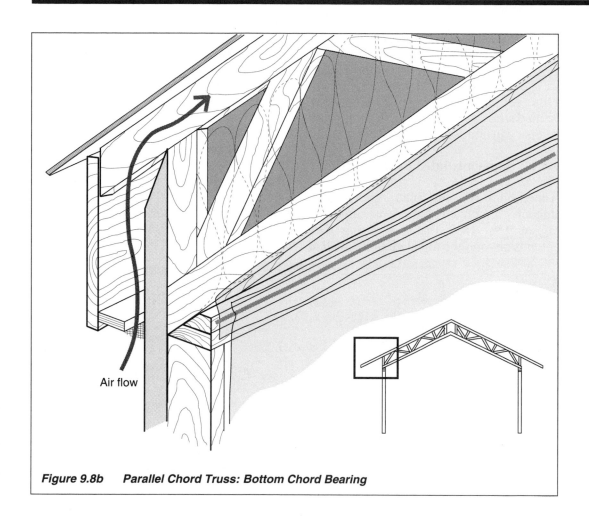

Air flow

Figure 9.8b   Parallel Chord Truss: Bottom Chord Bearing

9.3 SUPERINSULATED FRAMED ATTICS AND ROOFS

This section discusses ways of building framed attics and roofs to incorporate high levels of insulation.

9.3.1 Superinsulated Framed Attic

A method of framing an attic space using standard dimension lumber is shown in Figure 9.9. Note that the ceiling joist extends beyond the top plate of the wall at the eaves to produce a triangulated structure that can accommodate higher levels of insulation at the perimeter wall.

Advantages

• This method allows for higher insulation levels above the top plate.

Disadvantages

• The structure is more expensive to construct than a truss roof in most cases.

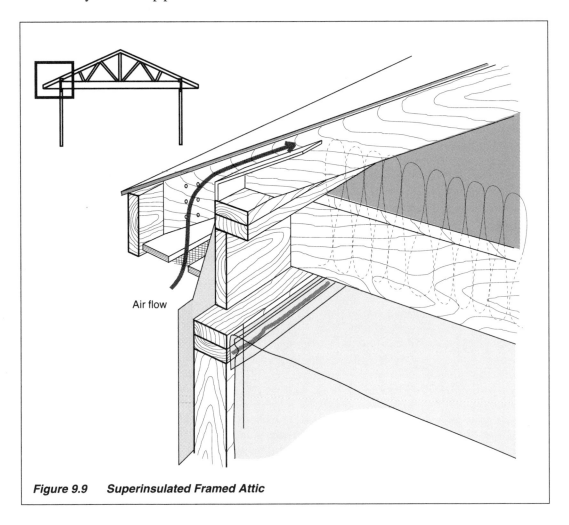

Air flow

Figure 9.9   Superinsulated Framed Attic

9.3.2 Framed Cathedral Ceiling

Figure 9.10 shows how to frame a cathedral ceiling using standard dimension lumber. Use rafters of 38x286 mm (2x12 in.) with cross-purlins of 38x38 mm (2x2 in.).

This framing allows for adequate ventilation of the roof space. Place two layers of RSI 3.5 (R-20) batts between the joists.

Note that the use of high-density RSI 7.0 (R-40) batts, which are 264 mm (10-3/8 in.) thick, eliminates the need for cross-purlins if they are not required by local building codes.

Advantages

- In some areas, costs can be lower than for parallel truss construction.

Disadvantages

- The clear span is limited by the maximum size of dimensional lumber to 38x286 mm (2x12 in.).

- Insulation is limited to a maximum of RSI 7.0 (R-40).

- Thermal bridging occurs through the rafter.

- There are extra material and labour costs.

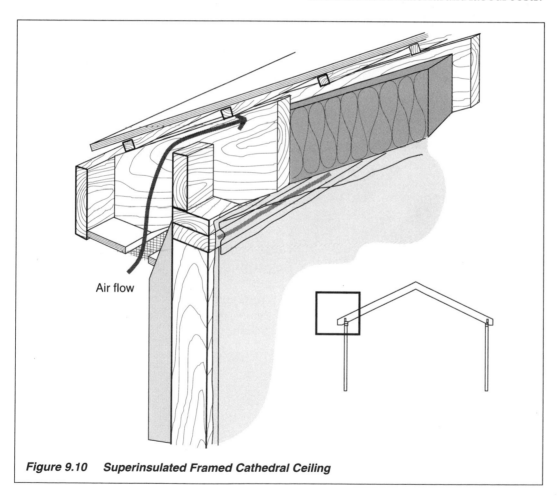

Air flow

Figure 9.10   Superinsulated Framed Cathedral Ceiling

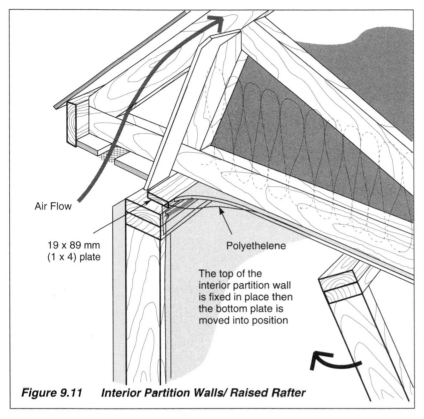

Air Flow

19 x 89 mm
(1 x 4) plate

Polyethelene

The top of the
interior partition wall
is fixed in place then
the bottom plate is
moved into position

Figure 9.11   Interior Partition Walls/ Raised Rafter

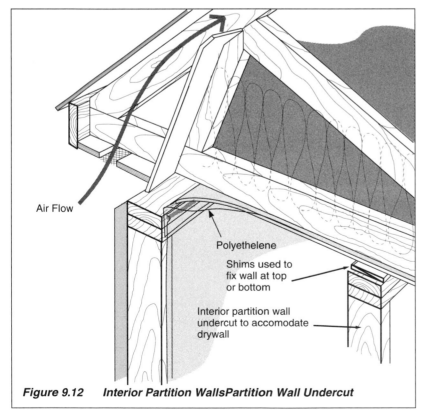

Air Flow

Polyethelene

Shims used to
fix wall at top
or bottom

Interior partition wall
undercut to accomodate
drywall

Figure 9.12   Interior Partition WallsPartition Wall Undercut

9.4   INTERIOR PARTITION/ CEILING JOINTS

Interior partition construction can greatly affect the energy efficiency of a ceiling, because considerable air can move through the ceiling/partition wall junction. Special care is required to keep the air barrier continuous.

9.4.1   Rigid Air Barrier Approach

There are two approaches to ensuring continuity of a rigid air barrier. One is to apply polyethylene and drywall across the entire clear span of the ceiling before the interior partitions are erected. Standard precut studs can be used for interior partitions if a 19x89 mm (1x4 in.) plate is placed on top of the exterior wall top plate.

This raises the trusses enough to accommodate 16 mm (5/8 in.) drywall and to tip the interior partition walls into place. The drywall can be taped before the interior partitions are raised (see Figure 9.11).

Another method is to undercut the interior partition walls by 25 mm (1 in.), erect them, and temporarily hold them in place until the drywall is "slipped" into place over top. They can then be anchored (see Figure 9.12). Alternatively, the studs can be undercut by 12.7 mm (1/2 in.) and the partition walls tipped into place.

Another approach is to seal the top plate of the partition wall to the wall and ceiling drywall.

9.4.2 Membrane Air Barrier Approach

If using polyethylene, ensure continuity of the air barrier by placing a 600 mm (24 in.) wide strip of 0.15 mm (6 mil) polyethylene between partition wall top plates. Seal the polyethylene flange to the ceiling air barrier in each room before installing the ceiling drywall (see Figure 9.13).

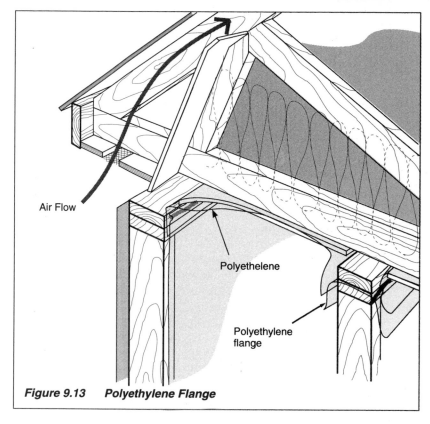

Air Flow

Polyethelene

Polyethylene flange

Figure 9.13   Polyethylene Flange

9.5   CEILING PENETRATIONS

Ceiling penetrations must be carefully sealed to minimize air flow into the attic space.

9.5.1   Plumbing Stacks

Wherever possible, minimize the number of plumbing stacks that penetrate the ceiling by bringing vents together inside the envelope.

Use an air and vapour tight flexible seal to prevent air leakage where plumbing stacks penetrate the ceiling at partition wall top plates. The seal assembly must be flexible, since plumbing stacks move because of expansion and contraction as well as shrinkage and settling of the house frame. Plumbing expansion joints are one way to minimize thermal expansion and contraction effects.

Figure 9.14 illustrates several methods of sealing plumbing stacks to the air barrier.

• One method is to pass the plumbing stack through a rubber gasket secured with a wood panel sheathing collar. The hole in the rubber is cut 40 mm (1-1/2 in.) smaller than the plumbing stack diameter and forms a tight friction fit when the stack is forced through it. Another approach is to cut an "X" in the gasket, keeping the cuts slightly shorter than the diameter of the pipe.

• The rubber gasket may also be secured by stapling into compressible foam tape as shown in the rigid air barrier approach in Figure 9.14.

• An "off-the-shelf" neoprene rubber roof jack will also work well and may be less expensive.

• An equally effective approach, generally used with rigid air barriers, is to seal the stack permanently and rigidly to the air barrier and to use a more flexible plumbing system. This will require an expansion joint in the stack, or an offset and a horizontal run of pipe close to the vent in the attic.

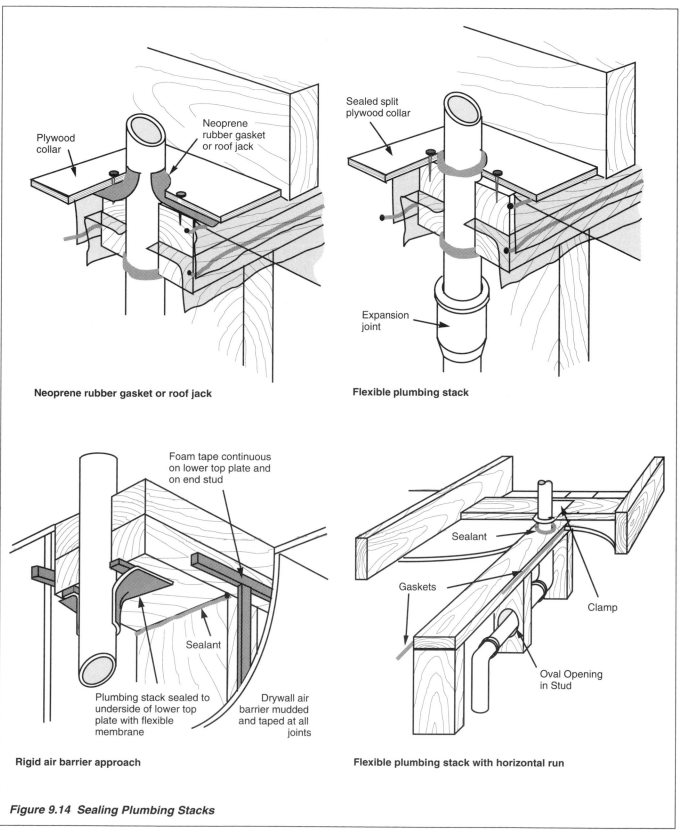

Neoprene rubber gasket or roof jack

Plywood collar

Neoprene rubber gasket or roof jack

Flexible plumbing stack

Sealed split plywood collar

Expansion joint

Rigid air barrier approach

Foam tape continuous on lower top plate and on end stud

Sealant

Plumbing stack sealed to underside of lower top plate with flexible membrane

Drywall air barrier mudded and taped at all joints

Flexible plumbing stack with horizontal run

Sealant

Gaskets

Clamp

Oval Opening in Stud

Figure 9.14 Sealing Plumbing Stacks

9.5.2   Wiring

Wiring can penetrate the ceiling through partition wall top plates. Seal penetrations directly to the air barrier and top plate with acoustical sealant.

9.5.3   Electrical Boxes

Electrical boxes for lighting fixtures can be sealed in several ways by:

- placing the electrical box in a prefabricated polyethylene envelope (see Figure 9.15)
- placing the box in a site-built wood box wrapped in 0.15 mm (6 mil) polyethylene
- placing the electrical box in a strapped cavity
- using a pancake surface-mounted electrical box (see Figure 9.16)
- bringing a single lead from the switch to the ceiling light fixtures, or
- using an airtight electrical box.

9.5.4   Recessed Lighting

Recessed lights are very popular, particularly in bathrooms and kitchens, both of which are areas of high moisture production.

Where recessed lights are required in the top storey ceiling, use "IC" CSA-rated fixtures which are designed to be in direct contact with insulation materials and which minimize air leakage through the fixture.

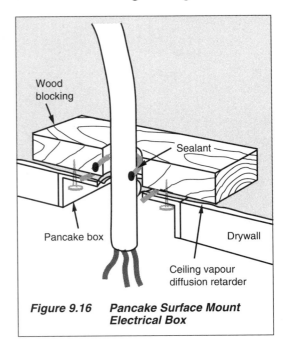

Figure 9.16   Pancake Surface Mount Electrical Box

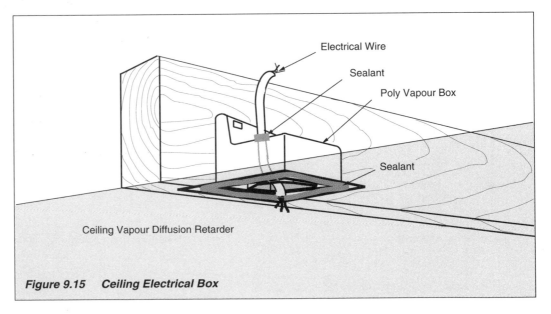

Figure 9.15   Ceiling Electrical Box

9.5.5 Attic Access Hatch

If the attic access hatch must pass through the ceiling, as in a hip roof, seal the hatch frame to the drywall and weatherstrip, insulate, and latch the access door (see Figure 9.17).

9.5.6 Ventilation and Heating Ducts

Air leakage from poorly sealed ducting running through unheated space is now acknowledged as a major factor in depressurizing houses, with the resulting increase in air leakage. Where distribution ducting must run into the attic space, all joints and seams in heating ducts must be taped.

Where ducts pass through the ceiling, they should be sealed to the air barrier with caulking (see Figure 9.18). Insulation must be placed over the ductwork to an RSI (R) value equal to that found elsewhere in the attic space. If the duct passes through the top plate of a partition wall, it must be sealed in the same manner as a plumbing stack.

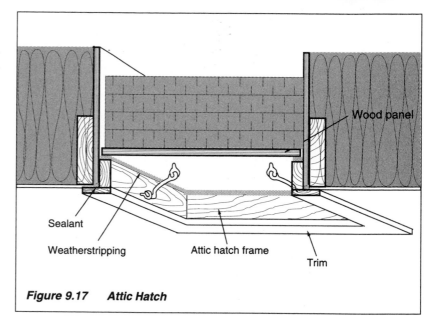

Figure 9.17   Attic Hatch

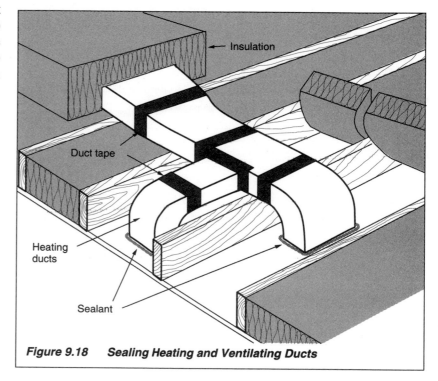

Figure 9.18   Sealing Heating and Ventilating Ducts

9.5.7 Chimneys and Vents

Figures 9.19 and 9.20 illustrate the procedure which should be used to seal around prefabricated metal chimneys. The same procedure should be used to seal the space around masonry chimneys. Figure 9.21 illustrates a method for sealing masonry chimneys when using a rigid air barrier.

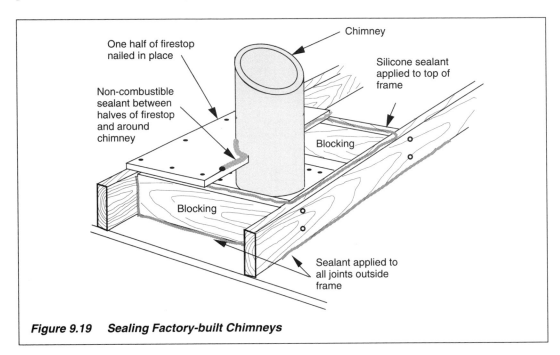

Figure 9.19   Sealing Factory-built Chimneys

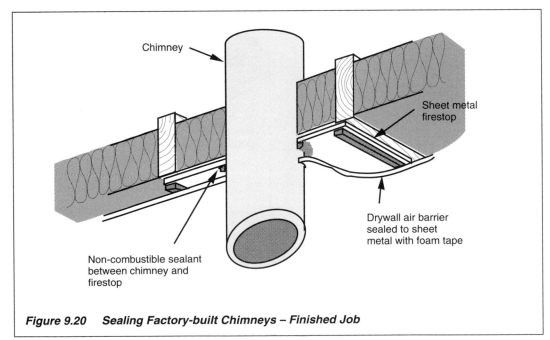

Figure 9.20   Sealing Factory-built Chimneys – Finished Job

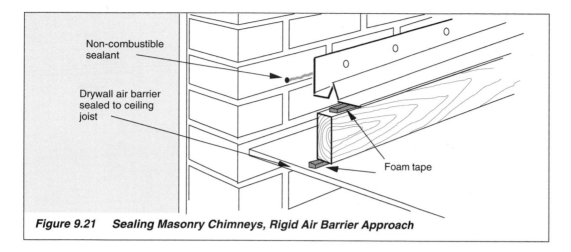

Non-combustible sealant

Drywall air barrier sealed to ceiling joist

Foam tape

Figure 9.21   Sealing Masonry Chimneys, Rigid Air Barrier Approach

9.6   NORTHERN CONSIDERATIONS

Most building codes require ventilation of the attic/roof space. Ventilation helps to exhaust any moisture or water vapour that may enter the space and keeps the attic/roof space cooler in summer. The most common technique is to use a continuous soffit ventilation strip and continuous ridge vents.

However, in northern regions, where fine blowing snow is common, local authorities may approve unvented attic/roof spaces.

9.6.1   "Hot Roofs"

A "hot roof" provides no ventilation in the roof or attic space. This type of roof assembly has been used successfully in the Northwest Territories where it evolved out of necessity. In Arctic areas with fine blowing snow, ventilated roofs have often not performed well. The Northwest Territories Housing Corporation has retrofitted the ventilated roofs on its existing stock of housing units to make them into hot roofs.

This type of roof should be used only if:

- the roof is tightly constructed to prevent the entry of fine, blowing snow
- heat loss through the ceiling is minimized by using proper air leakage control practices and an adequate layer of insulation which extends out over the top plates, and
- snow is not allowed to accumulate on the roof — either the wind must blow it off, requiring an average annual wind speed of 16 km/h (10 mph), or it must be removed by the occupants.

Close supervision and careful attention to detail are required if the hot roof approach is to be effective. However, keep in mind that this type of roof is not yet permitted by building codes in most areas of Canada and you would need a special waiver to build one.

Chapter *10*

MANUFACTURED HOUSING

Modular manufactured housing is an option that R-2000 builders may want to consider. The factory environment offers a number of advantages over on-site construction. For example, inclement weather will not affect construction schedules, and the centralized nature of the factory operation tends to make better use of materials.

Experience has shown that production line techniques can be adapted to accommodate the level of detailing demanded by the R-2000 Technical Requirements. In fact, the "inside out" fabrication of modular housing — where the air/vapour barrier and drywall are applied to the exterior walls before the mechanicals, electricals, insulation, sheathing and siding are installed — may provide for a more air tight structure. This sequencing allows air leakage in the exterior envelope to be readily identified and corrected during construction.

However, for the R-2000 builder, contracting to a factory builder means that control of the actual construction is in someone else's hands. Before turning over that control, the builder should be aware of the issues raised in this chapter.

10.1 R-2000 BUILDERS AND MANUFACTURED HOUSING

When considering contracting to a modular manufacturer, look for one who is certified by the Canadian Standards Association (CSA) and follows CSA-A277 when inspecting the performance of each work station. Ideally, the manufacturer should also be a certified R-2000 builder. This means its key personnel has undergone R-2000 training, and the firm has built an R-2000 demonstration home. "Key personnel" include:

- the quality control manager who is responsible for administering CSA inspections and ensuring that R-2000 Technical Requirements are met
- the quality control personnel who are on the plant floor at all times of production, and
- the engineering-design department members who are responsible for viewing all building plans to ensure they meet the R-2000 Technical Requirements; they also must be able to run the HOT-2000 Design Evaluation Program.

In addition, production personnel must be made aware of the need to pay strict attention to detail and to be familiar with the correct building techniques prescribed by the Technical Requirements.

Finally, you may need to consult with the plant purchasing department to make sure that they are aware of the materials which are, and are not, acceptable to the R-2000 program.

10.2 SPECIFICATIONS

The building design will have to be reviewed with the manufacturer's engineering department to ensure that it meets CSA standards, the manufacturer's standards, and the R-2000 Technical Requirements. HOT-2000 should be run at this time to ensure that the house design meets the R-2000 performance criteria.

The CSA-A277 standard requires that an inspection checklist be kept for each home produced — complete with details of checks performed at each work station. The R-2000 Technical Requirements should be integrated into this inspection procedure. Figure 10.1 is a comprehensive list of checks which must be included in the quality control system of the modular manufacturer in order to meet the R-2000 requirements. Builders should review production and inspection procedures to assure themselves that the technical requirements can, and will, be met.

In general, R-2000 construction techniques can be readily adapted to the production line. However, special attention should be paid to the integrity of the air/vapour barrier at the junctures where modules will be joined.

R-2000 Requirement	Checked By	Approved By
• Vapour barrier backing installed between joists		
• Subfloor sheathing glued and sealed at joints where required to meet R-2000 standards		
• Plywood spacer under exterior walls glued, sealed and fastened to subfloor, as required		
• Caulking or gasket installed between plywood spacer and sill plate of exterior walls		
• Shallow plenum ducts for HRV correctly installed and joints taped as required		
• Wall vapour barrier correctly fastened to wall and completely sealed		
• Wall vapour barrier correctly detailed at wall sill and top plate		
• All exterior wall electrical boxes correctly sealed		
• All plumbing pipes and mechanical ducting in exterior walls correctly sealed		
• All electrical wires penetrating through exterior wall sill plate correctly sealed		
• Ceiling system correctly attached and sealed to exterior wall system (check corners)		
• All interior walls correctly fastened to ceiling and sealed as required using plywood washers and caulking		
• All mechanical and plumbing perforations through ceiling vapour barrier resealed using accepted methods		
• Fireplaces correctly installed and sealed through vapour barriers and appropriate combustion air duct installed		
• Appropriate eaves baffles/foam insulation and caulking correctly installed		
• All HRV ducting installed in attic as required		
• Attic insulation installed as required and at the correct levels		
• Attic access hatch correctly fabricated and installed, complete with weather stripping and latches		
• Exterior HRV ducting vents installed in gables/soffit/roofs as specified		
• Weather barrier installed and taped as required (optional)		
• All windows and doors installed, insulated and sealed as R-2000 Technical Requirements		
• Gaskets correctly installed and sized at "marriage" joints, where modules fit together		

Figure 10.1   In-Plant R-2000 Quality Control Checklist

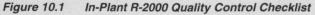

Source: *Canadian Modular Manufacturered Housing and R-2000: A Guide to Canadian Standards Association Quality Assurance, May, 1989*

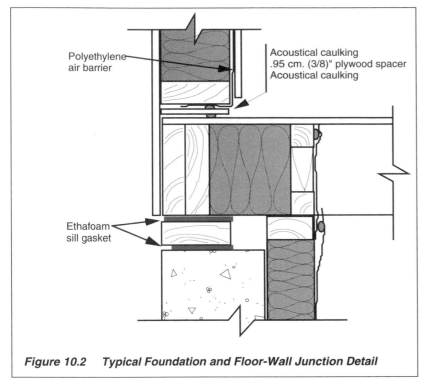

Figure 10.2   Typical Foundation and Floor-Wall Junction Detail

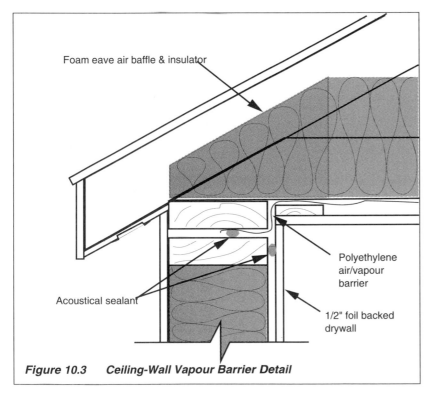

Figure 10.3   Ceiling-Wall Vapour Barrier Detail

Floors

The structural tie between the floor and exterior walls requires special consideration. The walls must be fastened securely to the perimeter beams of the floor system to withstand the rigours of lifting and transportation to the site. For the same reason, care must be given to obtaining the required air/vapour barrier detailing at this point. Figure 10.2 shows how some manufacturers are handling the installation of the air/vapour barrier and insulation in this area.

Wall-Ceiling Junction

Roofs are generally prefabricated and then lifted as one unit — with trusses, strapping, air/vapour barrier and drywall already attached — into place.

If a foil backed drywall is used in place of polyethylene, a suitable gasket glue or caulking material must be used to ensure a continuous seal between the drywall and the top plate (see Figure 10.3).

Marriage Joints

Perhaps the most difficult area to maintain the integrity of the air/vapour seal is where the modules join together. Modules are constructed to very accurate tolerances, and the space between them is very limited.

There are two distinct situations that need to be taken into account: single storey modules and multi-storey modules.

a) Single-storey Modules

In a single storey situation, the two areas that require the most attention are the ceiling joint and the exterior wall joint. Two gasket systems are commonly used. In one, a length of 0.15 mm (6 mil) poly formed into a long bag and filled with fibre batts —150 mm (6 in.) wide by 25 to 75 mm (1 to 3 in.) thick for wall sections and 300 to 400 mm (12 to 16 in.) wide by 25 to 75 mm (1 to 3 in.) thick for ceiling sections. These bags are then caulked and fastened to half of the modules.

The second system uses an ethafoam-type gasket material between the modules. The material must be thick enough to create a complete seal. Obviously great care must be taken during transportation and when setting the modules in place to ensure that the gaskets are not disturbed.

b) Multi-storey Modules

An additional layer of complexity is added in multi-storey situations, when modules must be stacked. A number of techniques have been used successfully. For example, ceilings have been installed on the lower modules as well as the floor system on the upper module; in effect, stacking two single storey modules. It is also possible to leave the floor off of the upper module, or, vice versa, the ceiling off of the lower module.

Pay strict attention to the air/vapour detailing along the cold edge of the perimeter of the marriage joints on each module (see Figure 10.4).

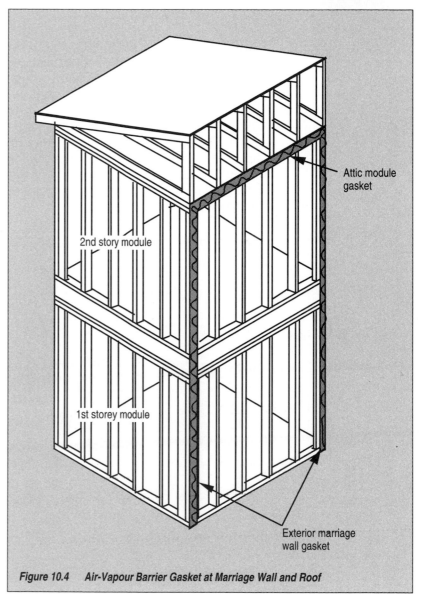

Attic module gasket

2nd story module

1st storey module

Exterior marriage wall gasket

Figure 10.4   Air-Vapour Barrier Gasket at Marriage Wall and Roof

10.3   DELIVERY

Arrangements can be made to have the manufacturer deliver the modules to the site. To make sure that the integrity of the structure and envelope is maintained, the builder should be on the site at the time of delivery to monitor the off-loading process . The units should be thoroughly inspected at this point. Special attention should be given to the inspection of any loose fill insulation, as well as the gaskets between the modules.

10.4   STITCHING

Perhaps the most crucial aspect of R-2000 housing constructed by a modular manufacturer is the "stitching" together of the modules. If the completed house is to pass the air leakage test, the integrity of the air/vapour barrier has to be maintained during this final phase. Figure 10.5 lists the areas that must be thoroughly inspected and signed off in accordance with the CSA and R-2000 standards.

Depending on the complexity of the design, it can take from one working day to several weeks to stitch the modules together. Tradespeople working on the various stitch tasks must be thoroughly trained and familiar with the technical requirements of the R-2000 program.

R-2000 Requirement	Checked By	Approved By
• List damage to module caused by transport and correct as required		
• Inspect attic and ceiling insulation to ensure no displacement during transport (correct as required)		
• Check all exterior module gaskets ensuring correct installation without voids		
• Ensure caulking applied to warm side of marriage joint on modules without gaskets		
• Monitor setting of modules on foundations ensuring gaskets between modules not damaged during setting		
• Ensure gaskets sufficiently compressed to create a continuous seal at all required locations		
• Inspect all foundation sill gaskets and plates to ensure adequate sealing and finishing		
• Inspect all work performed by on-site work crews related to R-2000 Technical Requirements and ensure it meets the required standards		
• Check all HRV duct joints to ensure they are correctly connected		
• Check HRV after completely installed to confirm it is correctly balanced and delivering sufficient cfm to each area of the home		
• Coordinate on-site air tightness testing of completed unit		
• Inspect exterior air barrier and ensure it is correctly installed before exterior siding is installed		

Figure 10.5   On-Site R-2000 Quality Control Checklist

Source: Canadian Modular Manufacturered Housing and R-2000: A Guide to Canadian Standards Association Quality Assurance, May, 1989

10.5   *MANUFACTURED HOUSING*

Over the last several decades, factory-based manufacturing has played an ever increasing role in the house construction process. Roof trusses, manufactured stairs, cabinets, framing materials and panelized wall systems have been widely adapted. Increasingly, factory built housing units (ready to move to the construction site) are gaining wider acceptance.

Manufactured housing cannot be considered a new phenomenon. Since even before the 1940s, Canadian manufacturers have produced complete houses or multiple units in the factory, ready for transport directly to the building site. As the years have progressed, these units have evolved from their "trailer-like" predecessors to fully customized units, adaptable to almost any design and consumer preference.

Those homeowners and builders who have used factory-manufactured products claim the following benefits:

• increased control over delivery schedules (greater control over the manufacturing process);
• reduced moisture-related problems (improved environmental control during the construction process);
• improved adaptability in the design process (computer assisted design equipment and processes); and
• reduced construction defects and Code infractions (improved construction management procedures).

More and more builders and developers are looking to manufactured housing to meet the needs of the home purchasing public. Manufactured units have been employed in single family projects, as the basis for entire subdivisions, and even in multi-storey buildings. Increased design and customization capabilities on the part of many manufacturers is allowing builders to meet the needs of any market segment.

10.5.1   Debunking the Myths

Resistance to factory-based housing continues to exist. Several myths continue to be widely held within the industry.

"Factory built housing is limited in design."

A common misconception is that factories produce housing units which have to look like trailers, or which must be produced in standard module sizes. A look at the variety of housing types produced in the factory puts this assumption to rest. Factory-based design and production can accommodate virtually any degree of customization.

"Factory built housing is of inferior quality."

A common misconception is that factory produced housing units are of a lower quality than stick built units. In reality, many of Canada's most energy efficient homes have been built in the factory and assembled on site. Sophisticated production and management practices allow for a high degree of quality control. Building in a controlled environment with materials which have been carefully stored minimizes many of the construction defects resulting from building in the Canadian climate. Quality control mechanisms and procedures are also easier to apply in the factory than on the site.

"Factory built housing won't meet consumer demands."

Most home purchasers are far more concerned with the final product than with the construction process. Where the benefits of factory-based production can be presented in relation to the delivered product, many customers will become willing buyers. Keep in mind that many purchasers must have those benefits demonstrated, either at the manufacturing plant or in a model home.

While it cannot be assumed that factory-based technology represents the wave of the future, it is clear that increased mechanization of the house construction process is gaining greater acceptance within the industry and the market. Builders who have not apprised themselves of the processes and products available through manufacturers of housing cannot adequately evaluate the future potential of these units. Only with enhanced awareness will builders make better decisions as to how factory-based units will affect their markets.

Chapter *11*

WINDOWS AND DOORS

The style and placement of windows and doors, particularly main entry doors and patio doors, is a major marketing consideration. At the same time, the number, type and placement of windows and doors in a house can significantly affect the building's heating load and occupant comfort. In typical housing, using conventional technologies, windows can account for as much as 35 per cent of the heat loss from an average house.

The past several years have witnessed major technological advancement in window technologies. Today's state-of-the-art windows provide energy performance which could only have been guessed at as little as ten years ago. Window advances have led to higher performance glazings with low emissivity coatings, gas fills and insulating spacers, and improved frames incorporating thermal breaks. If properly placed, high performance windows can now be viewed as an energy asset, rather than a liability. For example, most fixed high-performance windows will actually gain more heat over the heating season than would be lost. Design considerations for windows were discussed in Chapter 3.

The keys to reducing heat loss from windows and doors (and in optimizing solar gains) are to select good quality, energy-efficient units — and to ensure that installation minimizes air leakage around the outside of the frame. This chapter describes new window technologies and proper installation procedures for windows. The chapter also addresses issues relating to the purchase and installation of doors.

11.1   SELECTING WINDOWS

Heat Losses

Heat can be lost through windows by:

• conduction, convection and radiation through the insulated glazing (IG) unit itself
• conduction through the edge of the glazing unit (ie. spacer)
• air flow through operable components of the window and conduction through the frame, and
• conduction and air flow between the frame and the rough stud opening.

The first three sources of heat loss can be reduced by selecting appropriate units; the last can be reduced through proper installation (see Section 11.3).

Most window manufacturers have rated the thermal performance of windows not by thermal resistance (RSI / R value) but by the heat loss coefficient (U-value) — the rate at which heat transfers. The higher the U value, the faster heat flows through the material.

Of note, both RSI and U values are commonly published on the basis of the resistance or conductance only of the centre of the glazing. When the increased heat loss at the edges of the glass are factored into an equation, published values are almost always overstated in relation to the actual thermal performance of a given window.

Heat Gains

Windows can also serve as a major source of heat gain to the house, allowing solar energy into the occupied area. While in winter these gains are of major benefit — reducing heating requirements and providing respite from the cold outdoors — solar gains in the summer, and even at times in the spring and fall, can result in overheating.

Solar heat gains through windows are a function of the amount of available sunshine, the orientation of a given window and the Solar Heat Gain Coefficient (SHGC) of a given window. The SHGC is also called the shading co-efficient. It is expressed as the fraction of solar energy striking the window which passes through the glazing. The lower the shading coefficient, the lower the amount of sunshine which will enter the living space. For example, a window with a shading co-efficient of .89 transmits approximately 90% of available sunlight, while one with a shading co-efficient of .78 transmits just over 75% of available sunlight.

The hours of available solar energy will fluctuate and vary considerably throughout the country. Window orientation, as discussed in Chapter 3, will also affect the amount of solar gain and determine when those gains will occur — both over the course of the day and throughout the year. South facing glazing provides the greatest gains in the winter, and is most easily shaded against the higher sun during the warmer months.

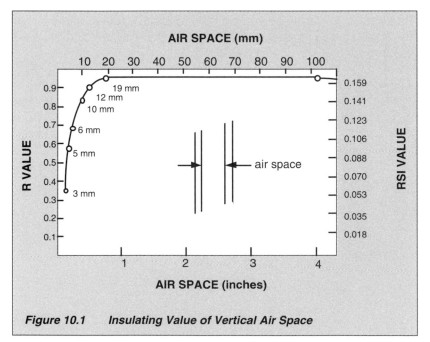

Figure 10.1   Insulating Value of Vertical Air Space

Several detailed computer programs are available for assessing the heat gain potential of different windows at different orientations and slopes in different climatic zones of the country. As a general rule, computer modelling of passive solar gains is not warranted in the design of the average house.

11.1.1 Trading off Heat Losses and Gains

The Energy Rating System

Identifying the actual performance of a window has historically rested within the domain of building researchers and designers. More recently, a standard test procedure to measure or calculate the heat loss and solar gain performance of an entire window or sliding glass doors has been adopted. This method, defined in CSA Standard A440.2, provides a rating on window performance based on:

- projected solar gains
- conductive, convective and radiant heat losses through the IG glazing
- heat conductive losses through the window frame, and
- heat losses associated with air leakage through the frame.

Windows can now be compared using a single number based on a standard sized unit in each of seven window categories. As a general rule, the performance of fixed windows will be better than other window types based on two factors. The overall area of the frame is smaller in relation to the glazed area of a fixed unit, and air leakage (through the unit) will be minimal. Standard sizes for measuring or calculating energy performance of different window types include:

- slider /tilt turn (920 x 1550)
- fixed (1220 x 1220)
- casement (600 x 1220)
- glass door (1830 x 2085)

The measured or calculated performance of a specific window is reported as the Energy Rating (ER). The ER is based on the <u>average</u> performance of windows at the four cardinal orientations, in a given climate, as measured or calculated over the heating season. The ER calculations are derived from the equation:

ER = solar gains - conduction losses - air leakage losses

A positive ER means that a window provides a net energy gain to the building; a negative ER means it loses energy. The ER can range from -80 for an operable window in a leaky metal frame with no thermal breaks, to +15 for a fixed window with high-performance triple glazing in a foam filled fibreglass frame. In comparison, a standard R20 wall would have an ER of -6.

Figure 11.2 compares ER values for several glazing and frame options of a typical casement window with a standard air leakage rating. The ER is useful in comparing two similar windows to determine which is more energy-efficient, but cannot be used to estimate the actual heat loss (or gain) through the window.

As stated, these ratings represent an average of performance of windows at different orientations. CSA- A440.2 provides information for calculating specific ratings (ERS) for windows at a specific orientation, in a specific location and for different house types (insulation levels & mass levels). Specific ratings based on a given orientation are available through most manufacturers and may prove useful in specifying window selection based on orientation. More sophisticated window choices may be provided through HOT-2000 modelling.

Glazing	Thermal Resistance RSI (R)	Shading Coefficient/SHGC
Single Glazing	0.15 (0.85)	0.87
Double Glazing: RSI(R) through the centre of the glazing		
12.7 mm (1/2 in.) air space	0.30 (1.70)	0.78
12.7 mm (1/2 in.) space filled with argon	0.37 (2.10)	0.78
12.7 mm (1/2 in.) air space, hard coat low-emissivity coating	0.47 (2.66)	0.75
12.7 mm (1/2 in.) air space, soft-coat low-emissivity coating	0.51 (2.90)	0.64
19 mm (3/4 in.) air space with low-emissivity coating between (i.e., two spaces of 9.5 mm (3/8 in.))	0.58 (3.29)	0.69
19 mm (3/4 in.) argon-filled space with low-emissivity coating between (two spaces of 9.5 mm (3/8 in.))	0.71 (4.03)	0.69
Triple Glazing - 12.7 mm (1/2 in.) air space	0.49 (2.78)	0.70

Window Type		
Double Glazing - 6 mm (1/4 in.) air space		
wood frame	0.31 (1.79)	0.78
metal frame	0.25 (1.41)	0.78
thermally broken metal frame	0.30 (1.69)	0.78
Double Glazing - 12.7 mm (1/2 in.) air space		
wood frame	0.38 (2.16)	0.78
metal frame	0.30 (1.69)	0.78
thermally broken metal frame	0.36 (2.04)	0.78
Double Glazing - 12.7 mm (1/2 in.) air space, low emissivity coating		
wood frame	0.49 (2.78)	0.64
metal frame	0.37 (2.10)	0.64
thermally broken metal frame	0.45 (2.53)	0.64
Triple Glazing		
wood frame	0.58 (3.30)	0.70
metal frame	0.41 (2.32)	0.70
thermally broken metal frame	0.51 (2.90)	0.70
thermally broken metal frame, low emissivity coating	0.76 (4.34)	0.44

Figure 11.2 *Thermal Resistances and Shading Coefficients for a Typical Casement Window with Standard Air Leakage Rating (Note Text, Previous Page)*

The Energy Rating has been formulated for residential buildings in a heating-dominated climate. Thus, it is appropriate for use as a comparative yardstick in Canadian housing but should not be used in assessing glazing options for commercial buildings or for cooling-dominated buildings. In the future, all windows may carry labels with their performance ratings under CSA A 440.2. This will give builders a better means to compare windows of a similar type.

The ER is one of many considerations in the purchase of a window. Appearance, durability and price will be equally important in many cases, except where a minimum ER level is specified in building codes.

11.1.2 Reducing Heat Loss Through the Glazing Unit

Manufacturers can minimize heat loss through window glazing by:

- optimizing the width of the air space between the panes to minimize convection
- increasing the number of still air spaces between panes to reduce conduction (see Figure 11.2)
- incorporating low-emissivity coatings on one or more of the layers of glazing to minimize radiant heat loss, and
- substituting inert low molecular weight gases for the air between panes.

High performance windows combine all of these features. Using available technology, centre glazing R Values can be increased from around R2 for a standard double-glazed unit to levels in excess of R12. Many of the high performance windows have cost paybacks from energy savings sufficient to pay for the increased capital cost several times over. High performance windows also offer the advantages of maintaining a warmer interior surface temperature — reducing the potential for condensation and enhancing occupant comfort. The following sections look at the components of high performance glazings.

Low-emissivity (Low-E) Coatings

Most of the heat lost through standard windows is radiant heat: the inside panel of glass absorbs heat from the room and radiates it to the cooler outside panel. This type of heat flow can be reduced with low-emissivity (low-E) coatings — thin metallic coatings that slow the rate at which the glass will radiate heat.

A low-E coating on one sheet of glass in a double-glazed window will give that window an insulating value approximately that of a standard triple-glazed window (see Figure 11.2). Double-glazed windows equipped with these coatings weigh significantly less than standard triple-glazed windows with comparable RSI (R) values. This is a particular advantage in northern areas, where high RSI (R) values are needed but transportation is expensive. The reduced weight of high-performance double-glazing also makes it easier to operate than a standard triple-glazed window. This means less wear and tear on the window hardware (hinges, casement cranks, etc.) and longer window life. Triple glazed, low-E windows employing a middle layer of film provide enhanced performance with lower weight.

Although low-emissivity coatings increase the R-value of a window by reducing radiant heat loss, they may also reduce the amount of solar heat gain. Figure 11.2 includes solar heat gain coefficients for various glazing configurations. The effect of low-E coatings is easy to see.

The selection of glazing type should take into account the orientation of the specific window. North facing windows with little opportunity for solar gain in winter, should be selected on the basis of maximum RSI (R) value. Lower heat gain coefficients will not represent a major disadvantage for north facing windows. On the other hand, south facing windows should be selected to optimize solar gains.

Most Canadian manufacturers are selecting 'northern' coatings for their glazings. This approach allows high solar energy transmission at only slightly reduced emissivity effectiveness. Look at trade-offs between solar gains and heat loss before choosing glazing options. The simplest way to evaluate the trade-offs is by looking at the Energy Rating which allows a builder to select the product with the best balance between reduced heat loss and increased solar gains.

Gas-filled Units

More recently manufacturers of high performance windows are replacing air in the space between layers of glazing with heavy, inert gases such as argon and krypton. These heavy gases discourage convection within the space, thereby reducing heat flow through the window. While krypton is the better performing gas, it is also significantly more expensive. Argon filled units now represent the more common approach to the assembly of high performance windows.

The procedures for filling the entire sealed units with the heavy gas without allowing air to 'infiltrate' the space are reasonably complicated. Since there is no standard, manufacturers use their own systems, with varying degrees of success. The Insulating Glass Manufacturers' Council of Canada (IGMAC) currently has a test for standard glazing units. That test is being modified to include argon gas retention.

This procedure will ensure, among other things, that the concentration of fill gas in the cavity is adequate to provide good thermal performance (more than 90 per cent fill gas is required).

Before leaving the discussion of centre-glass heat loss, it should be emphasized that most manufacturers only quote centre-glass R-value when comparing their windows to other products. Remember: you're not buying just the centre of the window, but the whole unit, so total-window R-value is a more appropriate comparison. An even better comparison can be made using the Energy Rating.

11.1.3 Reducing Conduction Through the Edge of the Glazing

Considerable heat transfer occurs at the edge of the insulated glass unit, where a spacer is in contact with each layer of glazing. Manufacturers of high performance windows reduce this heat loss by:

• using insulating spacers and sealants, and
• increasing the width of the air space in the sealed unit.

Spacers have traditionally been made of aluminum. Some manufacturers now offer windows incorporating plastic, silicone, glass fibre, and composite spacers. These are far more resistant to heat loss, so they help to eliminate both failure of the seal around the edge of the glazing and condensation.

Increasing the width of the air space in the sealed unit allows use of a wider spacer but also permits convection currents to form in the space between the glazings. As shown in Figure 11.1, the optimum spacing between glazings is approximately 16 mm (5/8 in.).

11.1.4 Reducing Conduction and Infiltration Through the Frame

Conduction

Heat loss by conduction through the window frame can be reduced by using a low-conductivity material such as wood, PVC, or fibreglass for the frame. If a metal-frame window must be used due to durability, availability, or cost considerations, it should incorporate an effective thermal break (see Figure 11.6). In general, the wider the thermal break in a metal-frame window, the more effective it will be.

PVC frames with insulated reinforcing elements such as wood or glass fibre are also suitable. Large PVC windows usually incorporate reinforcing elements in their frames and sashes to reduce deflection.

Metal reinforcement can substantially reduce the R-value of a window; if possible, avoid it. Good quality PVC windows will not be subject to air leakage resulting from thermal expansion/contraction cycling.

Some window manufacturers now produce a PVC or fibreglass frame with foam-filled frame cavities. The foam filling reduces convection within the extruded cavities of the frame, thereby increasing the R-value of the window. Fibreglass (pulltruded) frames provide the advantages of low-conductivity and a low profile, resulting in a reduced area over which heat loss can occur.

The best thermal performance in a window frame can be achieved by using a fibreglass frame with fibreglass or foam insulation in the frame cavities. Total-window RSI-values of more than 1.1 (R-6) are possible with these frames.

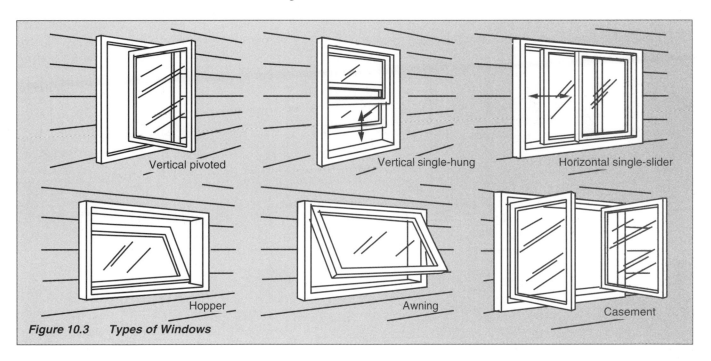

Vertical pivoted

Vertical single-hung

Horizontal single-slider

Hopper

Awning

Casement

Figure 10.3   Types of Windows

Infiltration

Infiltration through the frame is a major source of heat loss. Air leakage is often associated with increased condensation on the inside of the window. Use fixed (non-operable) windows whenever possible.

Operable windows should use durable, flexible gaskets to make an airtight seal between sashes and the frame. The airtightness of the joint depends on the type of weatherstripping used and the amount of pressure that can be applied by the closing system. Air leakage from sliding all of windows generally tends to be higher than from casement, awning, or hopper windows (see Figure 11.3), which can use compression seals. Warping of the frame or sashes also affects the airtightness of a window, as do changes due to thermal expansion and compression of components. Select windows with compression rather than sliding seals (see Figures 11.4a and 11.4b).

The A440 ratings of a window can be used as a rough guide to its construction quality. In general, a window can only achieve a high resistance to air, water, and wind leakage by having tight-fitting corner joints, good seals, and proper gaskets and weatherstripping. Air leakage characteristics of a given window are factored into the Energy Rating calculation.

Many building codes require windows to meet the A440 A1, B1 and C1 levels — levels based on varying design wind pressures. To minimize heat loss due to air leakage (particularly in windy locations or in high-rise buildings), select windows with an A3 rating.

In northern areas, consider durability and operability in extreme conditions. Opening windows in very cold conditions can cause a heavy accumulation of ice from condensing interior air making it difficult to close the windows again. Operable windows in northern houses should be extremely sturdy.

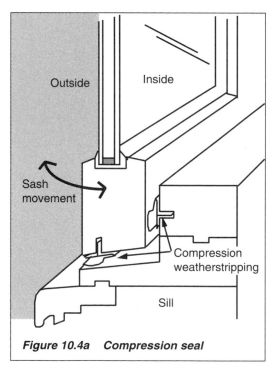

Figure 10.4a   Compression seal

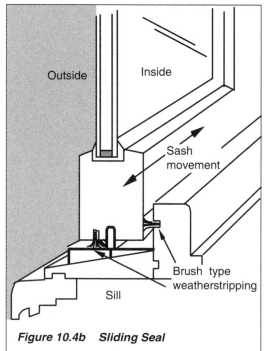

Figure 10.4b   Sliding Seal

11.1.5 Enhancing Solar Gain

The selection of windows for the home should include consideration of useful solar gains — those gains which contribute to the home during the heating season. As previously explained, many glazing technologies which reduce heat losses from the home can also reduce the useful solar gain by decreasing the solar heat gain coefficient (SHGC). Builders must consider window orientation as part of their decision making.

Several factors should be considered by the builder wishing to optimize solar gains for the house:

- Choose units with the best SHGC for south facing windows
- Multiple glazing layers and certain types of coatings will reduce SHGC — but may offer better RSI values required for north facing windows
- Choose windows with maximum clear opening — minimizing mullions, dividers etc.; and
- Minimize the size of the window frame and depth of profile to reduce shading.
- Design window placement and landscaping for best performance year-round (see section 3.3.1).

11.1.6 Summary

High performance windows are available from a wide range of manufacturers. The advantages of high performance windows extend beyond energy efficiency and savings to the homeowner. These windows also offer enhanced comfort and reduced condensation.

The Energy Rating system represents the builder's best guide to energy performance. Remember that the rating is developed to represent average performance over different orientations. The higher the ER, the better the performance (compare similar windows).

WINDOW RATING	Maximum air leakage rate $(m^3/h)m^{-1}$
A1	2.79
A2	1.65
A3	0.55
Fixed	0.25
	Water leakage test pressure differential (Pa)
B1	137
B2	200
B3	250
	Wind load resistance test pressure (kPa)
C1	1.5
C2	2.0
C3	2.5

*Figure 11.5   Minimum Window Performance Requirements :
CSA Standard A440*

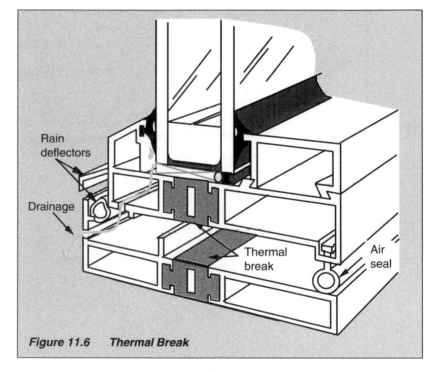

Figure 11.6   Thermal Break

Builders might also consider selecting different product features based on the orientation of the window. Windows with higher shading coefficients should be selected for the southern exposures, while windows on the north should be chosen on the basis of highest heat retention.

11.2 SELECTING DOORS

Reducing house heat loss through doors starts at the design stage. Try to:

- limit the number of doors in the house
- place doors out of the path of prevailing winds, either by locating them on the leeward side of the house or by providing windbreaks
- limit the number of patio doors or select efficient units, and
- use air-lock vestibules.

Then, as with windows, choose good quality units and install them properly. Heat is lost through doors by:

- conduction through the door and frame
- air flow between the door, frame and sill
- conduction through the glass in patio doors or in doors with lites, and
- air flow between the door frame and the rough frame opening.

The first three sources of heat loss can be reduced by selecting appropriate units; the last can be reduced through proper installation, as discussed in Section 11.3.

11.2.1 Reducing Conduction Through the Door

Much of the heat loss through the door assembly is lost by conduction through the door itself. Doors are made from a variety of materials, some of which reduce heat flow better than others. Solid wood doors will have insulating values ranging from RSI 0.35 to RSI 0.44 (R-2 to R-2.5). Metal-clad, insulated doors can have insulating values of up to RSI 2.46 (R-14), depending on the style of the door and the insulation material used to fill it.

11.2.2 Reducing Air Flow Between the Frame and the Door

Heat is also lost by air flow between the door and the frame. This heat loss may be reduced by providing an airtight seal between the door and the frame. The airtightness of the joint between the door and the frame depends on the type of weatherstripping used and the amount of pressure that can be applied on the joint. Select doors with compression seals rather than sliding seals; consider using insulated, prehung entry door systems which have good air seals (see Section 11.2.4).

11.2.3 Reducing Transmission Through Glazing

Heat can be lost through the glazing of door lites just as it can through windows. Refer back to Section 11.1.1 for tips on selecting glazing.

11.2.4 Selecting Patio Doors

In terms of heat loss and heat gain, patio doors act like large horizontally sliding or casement windows. Selection of patio doors should reflect an understanding of window selection procedures based on glazing performance. The Energy Rating System applies to sliding glass patio doors, so selection can be based on a comparison of ER ratings. Select units with good air seals. French doors and airplane-type patio doors provide a more positive seal against air leakage than sliding units.

11.2.5 Insulated Pre-hung Entry Doors

Insulated, pre-hung entry door systems provide better thermal performance than conventional wooden doors. These systems consist of an entry door with a core of polyurethane or polystyrene foam insulation. In the factory, the door is installed in the frame and sill system.

These pre-hung insulated door systems have several advantages:

- they have higher insulating values — up to RSI 2.46 (R-14) — than conventional wooden doors
- their air seals are tighter and more durable
- they come with a thermally broken adjustable sill assembly to reduce conductive heat loss, and
- no assembly time is required and installation time is less than that needed for site-assembled systems.

Pre-hung insulated door systems are available with steel- or wood-faced doors. A variety of materials may be used for the door framing, the insulation in the core and the weatherstripping. When selecting a system, consider:

- the thermal resistance of the door
- how the manufacturer prevents thermal bridging at the edge of the door and at the frame and/or sill, and
- tested air leakage rates for the door system in litres per second per metre of crack length (cubic feet per minute per foot of crack length).

An energy rating system for determining the performance of doors is not currently in place. It is likely that over the coming years a system comparable to the ER system will be instituted.

11.2.6 Summary

In selecting doors, look for:

- well-insulated cores
- wood, or thermally broken metal or PVC frames
- weatherstripping fabricated from high-performance, durable materials
- low air leakage rates (for pre-hung door systems), and
- maintenance-free framing materials.

11.3 WINDOW AND DOOR INSTALLATION

As shown in Figure 11.7, there are two ways to place windows and doors in relatively thick walls: close to the outside face of the wall, or close to the inside face. Doors tend to be mounted flush with the inside face for aesthetic reasons and for ease of operation.

It is generally more energy-efficient to mount windows close to the inside because:

* the window surface is sheltered somewhat from the wind, and
* condensation on the inside surface of the glass is reduced because air flow over the surface is improved and the window is in the warmer part of the wall.

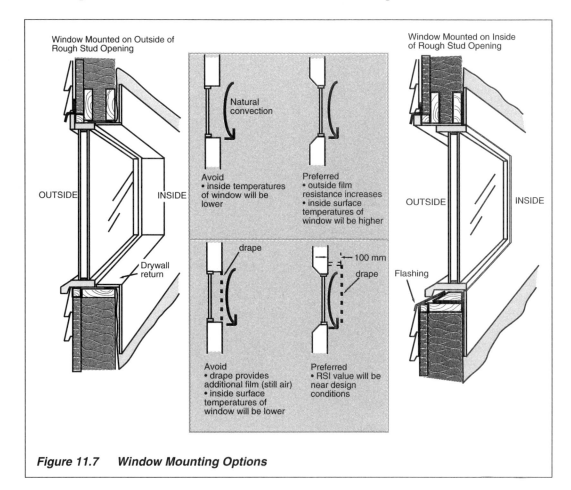

Figure 11.7   Window Mounting Options

However, inside mounting requires extra care to construct the deep weatherproof sill. It is more common to install the window flush with the exterior face of the wall with a recess on the inside and build out the frame or face the recess with drywall. It is also possible to use wood panel sheathing jamb extensions that are sealed to the outside of the window frame.

Windows and doors are typically installed in a space that is 25 mm (1 in.) larger than the frame in both the horizontal and vertical directions. Heat is lost through this "shim space" by conduction through whatever material has been used to fill the space and by air flow through the space.

To minimize this heat loss, the shim space must be insulated and sealed. The traditional way to fill the shim space has been to stuff scraps of batt-type insulation into the gap before applying the trim. This provides insulation but does little to reduce air flow through the space. The space must also be sealed to cut air flow around the window frame.

Some builders use the continuous polyethylene approach. This involves sealing a strip of polyethylene to the edge of the window frame and then sealing it back to the polyethylene sheet covering the wall.

The more common approach to air sealing involves filling the space with foam. Foams have a higher insulating value per unit of thickness than batt-type insulation and eliminate air flow more effectively. Care must be taken not to overfill the space, as the pressure from the expanding foam could warp the frame, making operation more difficult.

Another approach involves the use of contractor tape to seal the air barrier to the inside face of the window frame. Tapes must be compatible with both surfaces.

Finally, windows can be set and sealed into a plywood surround to minimize air leakage.

Detailed descriptions of several air sealing techniques for window or door frame into a wall are provided on the following pages.

11.3.1   Polyethylene Wrap

This method involves applying a 0.15 mm (6 mil) polyethylene flap to the window or door frame (see Figure 11.8).

A) Cut a 600 mm (24 in.) wide strip of 0.15 mm (6 mil) polyethylene. It should be long enough to go around the frame of the window or door with about 900 mm (36 in.) extra to allow for corner folds.

B) Apply a bead of acoustical sealant to one side of the frame, towards the inside (following the "one-third/two-thirds rule") so that condensation does not form on or around the frame. Seal all joints between the frame and any jamb extensions.

C) Lay the polyethylene strip over the bead of sealant, placing a pleat in the strip on both sides of each corner. The pleats allow the flap to be folded back against the polyethylene sheeting covering the walls. Staple the polyethylene to the frame through the sealant and inject sealant into the pleats. Continue this process around the entire frame and join the polyethylene to itself with acoustical sealant and staples. Use only enough staples to hold the polyethylene in place.

D) Place glass-fibre-reinforced tape over the caulking bead and staple through the tape to prevent the polyethylene from pulling off the frame during installation.

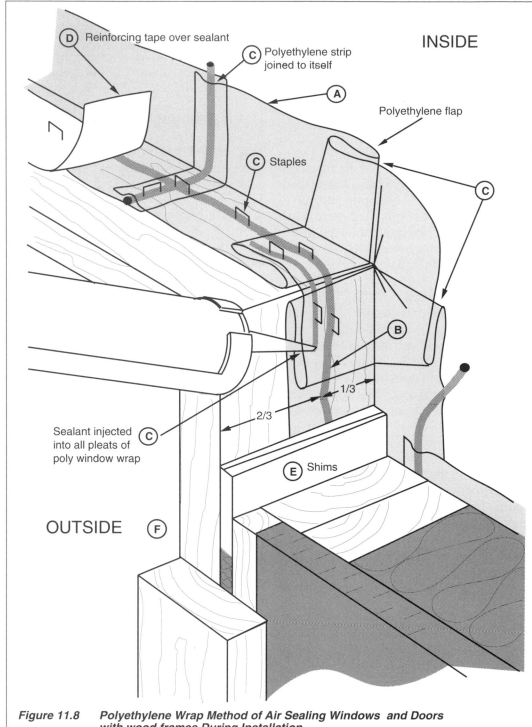

INSIDE

(D) Reinforcing tape over sealant

(C) Polyethylene strip joined to itself

(A)

Polyethylene flap

(C) Staples

(C)

(B)

1/3

2/3

Sealant injected into all pleats of poly window wrap (C)

(E) Shims

OUTSIDE (F)

Figure 11.8   Polyethylene Wrap Method of Air Sealing Windows and Doors with wood frames During Installation

E) Insert the window or door into the rough frame opening and shim it into place if necessary. Install wedges between the flap and the opening and not between the flap and the frame.

F) Fill the shim space with scraps of batt-type insulation or backer rod.

G) After applying the wall polyethylene, cut out the window and door openings. Apply a bead of acoustical sealant between the wall polyethylene and the window and door polyethylene flaps. Staple the two together through the bead of sealant.

11.3.2   Foam/Tape Method

This method can be employed on any window type. Complete filling of the void between the window frame and the rough stud opening is required. However, over-filling must be avoided.

- Fill the gap between the stud and window frame, allowing excess foam to 'bleed' to the interior.
- Trim the foam to the inside face of the window jamb.
- Tape or seal the polyethylene to the inside face of the jamb, prior to installation of trim.

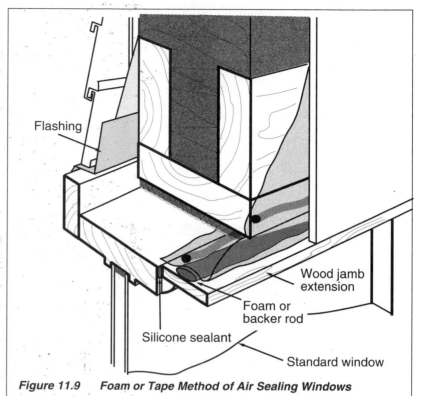

Flashing

Wood jamb extension

Foam or backer rod

Silicone sealant

Standard window

Figure 11.9   Foam or Tape Method of Air Sealing Windows

11.3.3   Plywood Wrap

This method is most commonly applied to metal and PVC windows and doors.

A) Frame the rough frame opening to accommodate a 12.7 mm (1/2 in.) wood panel sheathing liner covering the full depth of the opening on all four sides. This means an increase of about 25 mm (1 in.) in both the height and width of the opening.

B) Nail the liner into place flush with the interior finish and the exterior sheathing. The liner can be caulked to the rough frame opening on the interior.

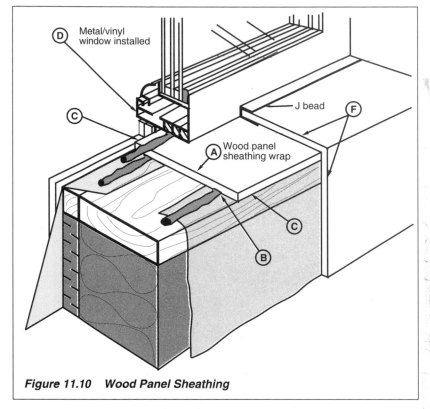

Figure 11.10   Wood Panel Sheathing

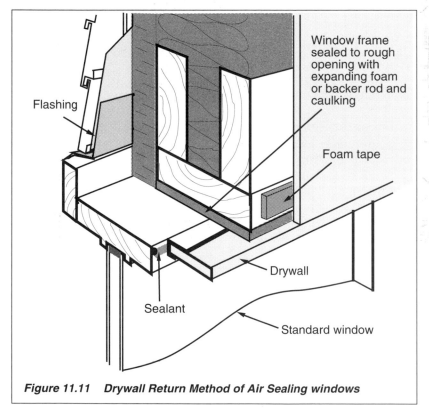

Figure 11.11   Drywall Return Method of Air Sealing windows

C) Seal the wood panel sheathing liner at the interior edge to the wall with either polyethylene or drywall (see Figure 11.10).

D) Install the window into the liner from inside or outside, depending on the intended location. If the window is to be mounted to the inside of the wall, slope and flash the sill properly prior to installation.

E) Insulate and seal the shim space between the wood panel sheathing liner and the rough opening.

F) Install drywall over the liner.

G) Install the finished sill (see Figure 11.10).

11.3.4   Drywall Method

This method is commonly used with wood-framed windows. The window is installed in the opening and the opening is finished with drywall sealed to the window frame (see Figure 11.11).

*   Frame the rough stud opening in the usual way.
*   Install window, insulate and seal the shim shape
*   Butt and seal the drywall to the window frame.
*   If the window is installed towards the outside of the wall assembly, install a drywall return in the recess. Caulk the joint between the face edge of the window frame and the drywall.
*   If the window is installed towards the inside of the wall assembly, the face of the drywall should be flush with the face edge of the window frame. Tape the joint and cover it with trim.

Part 3
MECHANICAL SYSTEMS

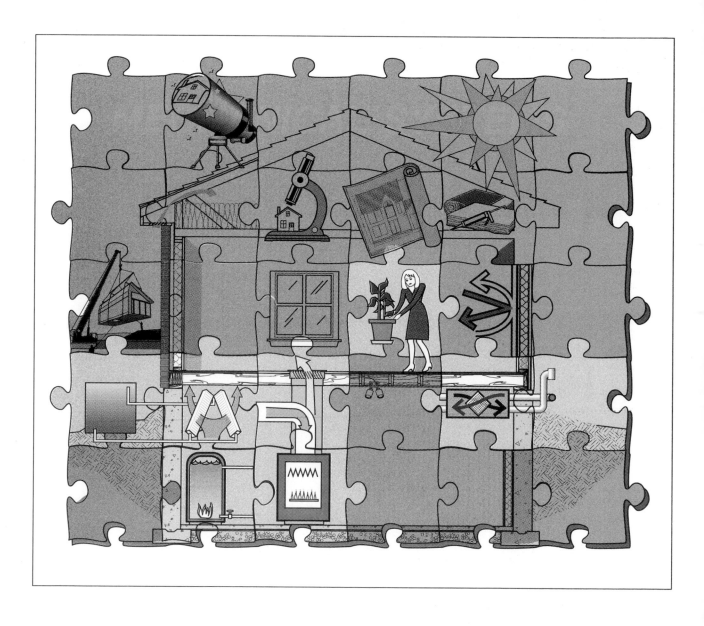

Chapter *12*

PRINCIPLES OF
SPACE CONDITIONING

Space conditioning is a general term used to describe the processes by which the air in a house is "conditioned" to make a comfortable living environment. Space conditioning includes:

- heating in winter
- cooling in summer
- ventilation
- the control of indoor relative humidity levels, and
- the filtration of air to remove impurities.

Efficient space conditioning equipment and systems are one of the hallmarks of a well built home. The design and installation of space conditioning systems is particularly important in homes with well sealed envelopes. This includes all homes built in accordance with the 1990 National Building Code. As was discussed in Chapter 1, the house operates as a system in which the space conditioning systems and the building envelope interact. This interaction has significant implications for both the comfort and health of the occupants and the energy consumption of the home.

This chapter reviews principles that are critical to the design and installation of safe, efficient space conditioning systems. These principles include requirements for combustion, dilution and make-up air, and general principles of air cooling, ventilation and filtration. Equipment and systems considerations for the various space conditioning processes are discussed in Chapters 13 to 18.

12.1 HEATING PRINCIPLES

Conventional fuel-fired furnaces and boilers require air for two purposes. They use the air's oxygen in the combustion process, and they use air to dilute and exhaust the products of combustion. In a typical house, combustion air accounts for about 1.5% of the heating load of the house over the heating season. Dilution air, on the other hand, is often a much larger and more variable quantity, and can account for 10-15% of the heat loss in a house.

In the past, heating systems in leaky houses drew on room air to provide combustion and dilution air. Outside air would leak in through random cracks and holes in the building envelope to replace this room air. However, this approach is not enough in today's houses with well-sealed envelopes.

Most of the new higher efficiency furnaces and boilers effectively eliminate the need for dilution air, as do the new high technology space heaters, such as the advanced combustion woodstoves and fireplaces.

12.1.1 Combustion Air

In a house built according to the techniques described in this manual, all combustion appliances must be provided with a supply of combustion air brought in from the outside. In some cases, the combustion air supply is integrated into the equipment design. If it is not, the source of combustion air should be sized and located according to whichever of the following codes is appropriate:

- CAN/CGA-B149.1-M93, *Installation Code for Natural Gas Burning Appliances and Equipment*
- CAN/CGA-B149.2-M93, *Installation Code for Propane Burning Appliances and Equipment*
- CSA B139-M91, *Installation Code for Oil Burning Equipment,* and/or
- CSA B365-M91, *Installation Code for Wood Burning Equipment.*

Combustion air ducting should be insulated and provided with an external vapour diffusion retarder (VDR) to avoid condensation on the outside duct surface. Condensation can also occur inside the duct if inside air flows back into the duct. Some designers put a damper in the duct to stop this backflow.

Details of wall penetration are important. Unless the VDR of the duct is sealed to the wall VDR where the duct penetrates the wall, the air leakage around the cold duct can create condensation problems in the wall. If a separate air barrier is used, the duct must also be sealed to it.

The same combustion air supply can serve two gas- or oil-fired appliances, as long as the appliances are located in close proximity to each other and the combustion air supply is sized to handle the combined requirements of both appliances. However, each wood-fired appliance must have its own combustion air supply, as described in the next section.

12.1.2 Make-up Air

Make-up air is required in houses where exhaust appliances induce a negative pressure which might lead to the backdrafting and spillage of combustion products from fuel-fired appliances. While upgraded standards have resulted in a phasing out of naturally aspirating heating equipment — being replaced by induced draft and condensing units — spillage can still be of concern where homes are equipped with wood burning fireplaces, wood stoves and similar appliances.

12.1.2 Wood Burning Appliances

Special attention must be paid to wood-burning appliances in airtight homes. Fireplaces and conventional wood stoves should be discouraged as they can have adverse effects on both energy consumption and air quality. However, many homeowners are willing to put up with the potential problems for that warm, homey feeling, and builders must be willing to meet their requirements in the safest possible manner.

Wood burning appliances have improved greatly in the past several years, both in their efficiencies and in their pollution control mechanisms. New appliances generally consume considerably less air for the combustion process. At the same time, the potential for backdrafting exists. A supply of fresh outside air to the appliance should be considered.

Chapter 14 provides greater detail on the design and installation of both fireplaces and wood burning appliances. Canadian Standards Association (CSA) and the United States' Environmental Protection Agency (EPA) certification may be required in some jurisdictions.

12.2 COOLING PRINCIPLES

Space cooling or "air conditioning" is often necessary during warmer months to remove heat resulting from:

- solar heat gain
- heat gain through the components of the building envelope
- air infiltration
- air brought into the house for ventilation
- appliances and lights that produce heat
- activities of the occupants such as showering or cooking, and
- the occupants themselves — an adult at rest gives off 120W of heat.

Cooling equipment also provides dehumidification by condensing water vapour out of the air as it is chilled.

The cooling load is calculated in the same way as the heating load (see Chapter 2), but in reverse: the concern is with heat gain in the occupied space. Heat gain will be a function of:

- the size, orientation and shading of the windows and patio doors
- the envelope insulation levels
- the air leakage rate of the structure, and
- the ventilation rate and efficiency of a heat recovery ventilator (HRV)
- the impact on the micro-climate of shade trees and other landscape features
- the adequacy of attic venting - summer temperatures can reach 65°C (150°F).

As with heating equipment, the cooling unit must be carefully sized. Undersized systems will not provide adequate comfort under design conditions. Oversized systems will be unnecessarily expensive and will not provide adequate humidity control. The design of distribution systems for cool air is also very important. Larger ducts may be needed because cooling systems usually move more air than heating systems, and cooling produces heavier air which is more difficult to move.

12.3 VENTILATION PRINCIPLES

Ventilation is the process of removing indoor air and replacing it with fresh air from the outside. This two-way air flow is necessary to remove indoor pollutants and to provide occupants of the house with a healthy indoor environment.

As was noted in Chapter 3, a wide range of pollutants can be generated within the living space. As well, relative humidity levels can be very high when the house is new and still drying out. Humidity can also become excessive during periods of high occupancy, while meals are being prepared or when members of the family bathe or shower. Ventilation must be provided to control the concentrations of water vapour and pollutants in the air.

Additional ventilation is also necessary to replace air used by air-exhausting appliances.

These include clothes dryers, bathroom fans, indoor barbecue grills, fireplaces and central vacuums, in addition to combustion appliances. Figure 12.1 provides air flow rates for typical exhaust appliances.

In a house built with a well sealed envelope, a combination of these appliances operating together (or some even individually, like a wood fireplace or a barbecue grill) could exhaust enough air from the living space to cause a negative pressure. Such a negative pressure could also cause a smouldering wood fire to backdraft.

Backdrafting must be avoided by providing make-up air to these appliances; the air they exhaust needs to be replaced to balance the interior and exterior air pressures. In the case of a fireplace, the best solution is to use an advanced combustion design, as described in Chapter 14.

Make-up air can be supplied through a single intake or through individual inlets in each room. It should be provided in a manner that does not create discomfort for the occupants. For example, it may be possible to install a damper that opens only when the operation of air-exhausting appliances creates a significant negative pressure within the house. The advantage of using a single make-up air supply is that the air can easily be heated or cooled, if necessary.

However, supplying make-up air is only part of providing proper ventilation. You must also install a whole-house mechanical ventilation system to exhaust stale air and pollutants and to provide fresh air to all rooms in the house.

| Exhaust Device | Range of Air Flows | |
	litres per second (l/s)	cubic feet per minute (cfm)
Bathroom fan	20 - 50	40 - 100
Range hood	25 - 125	50 - 250
Indoor barbeque grille	60 - 150	120 - 300
Clothes dryer	40 - 55	80 - 110
Central vacuum	45 - 65	90 - 130
Fireplace (at full burn)	150+	300+

Figure 12.1   Air Flow Rates for Exhaust Devices

12.3.2   Ventilation and Building Standards

The need for mechanical ventilation to maintain a healthy indoor environment is increasingly recognized in Canadian building codes. For some time an exhaust fan has been required in electrically heated houses to reduce moisture. The fan exhausts air which would otherwise have been removed from the house by a chimney. A discussion of other developments in building codes and standards follows.

National Building Code

The *1990 National Building Code of Canada* adopted the following clauses requiring mechanical ventilation systems in all residential buildings.

9.32.3.1 Required Mechanical Ventilation for Dwelling Units

(1) Every dwelling unit shall be provided with a mechanical ventilation system having a capacity to exhaust inside air or to introduce outside air at the rate of not less than 0.3 air changes per hour averaged over any 24-hour period.

(2) The rate of air change required in Sentence (1) shall be based on the total interior volume of all storeys including the basement, but excluding any attached or built-in garage or unheated crawl space.

Most provinces have incorporated these clauses into their building codes. Others have implemented their own requirements which must be adhered to in the design of the ventilation systems. Upcoming building codes are currently expanding on these requirements.

Canadian Standards Association

A committee of the Canadian Standards Association (CSA) has been addressing the issue of ventilation in new residential buildings. CAN/CSA-F326-M91, *Residential Mechanical Ventilation Systems* incorporates a requirement for a continuous level of ventilation in a house, based on the number and types of rooms in the building. Two of the essential components of this standard are as follows:

- Continuous ventilation must be supplied at the rate of 5 litres/second (10 cfm) for each room, and 10 litres/second (20 cfm) for the master bedroom and the basement (see Figure 12.2). Exhaust capabilities must be supplied in the kitchen and bathrooms; a higher value of exhaust is necessary if continuous exhaust is not supplied to these rooms.

- Make-up air must be supplied to the house to offset pressure imbalances in excess of 10 Pa (0.04 inches of water) on a continuous basis. It allows variable intermittent pressure imbalances depending on the types of fuel-burning appliances installed in the house. The unbalanced air flow is determined by adding the unbalanced air flow of the ventilating equipment in its continuous-operation mode to the exhaust capacity of the single largest exhaust appliance in the house, plus an assumed clothes dryer exhaust capability of 75 litres/second (150 cfm).

Master bedroom	10 l/s (20 cfm)
Bedroom	5 l/s (10 cfm)
Bedroom	5 l/s (10 cfm)
Bathroom	5 l/s (10 cfm)
Kitchen	5 l/s (10 cfm)
Living room	5 l/s (10 cfm)
Dining room	5 l/s (10 cfm)
Family room	5 l/s (10 cfm)
Utility room	5 l/s (10 cfm)
Bathroom	5 l/s (10 cfm)
Basement	10 l/s (20 cfm)
Total continuous ventilation	**65 l/s (130 cfm)**
Total system capacity	**65 l/s (130 cfm)**

Figure 12.2 Ventilation Rates

12.4   RELATIVE HUMIDITY

As noted in Chapter 2, relative humidity is the amount of moisture in the air relative to the amount of water the air can hold at that temperature. When the relative humidity reaches 100%, the air is saturated. Any additional water vapour will condense in the form of water, frost or ice, depending on the temperature of the condensing surface.

Cold air holds less water vapour than warm air because the air molecules are closer together. When cold outside air leaks into a house during the winter or is brought in to provide ventilation and is heated to room temperature, the relative humidity in the conditioned space will drop. If it becomes too low, the air will draw moisture out of the furniture, carpets, walls and ceilings, and the occupants may complain of dry throats and static electricity. On the other hand, if the humidity becomes excessive, i.e. above 70%, moulds and mildew can flourish and wood can begin to rot (see Section 2.6).

12.4.1   Effective Temperature

Tests have been conducted for years to determine the combination of temperature, humidity and air velocity at which people are most comfortable. The "effective temperature", an index chosen to represent a combination of these effects, is used to measure the sensation of warmth or cold felt by the human body. It is defined as the temperature of saturated air (air at 100% humidity) that provides a sense of physical comfort when the air velocity in the conditioned space is between 7 and 12 metres per second (15 and 25 mph). Various combinations of temperature and relative humidity can provide the same sense of comfort (see Figure 12.3).

Space cooling systems can be used to provide dehumidification in summer. Equipment that can be used to add humidity to the air in winter is discussed in Chapter 13.

12.5   AIR FILTRATION

In general, the levels of dust and other particulates will be low in a well sealed home equipped with a properly designed mechanical ventilation system. Most heat recovery ventilators, for example, come with built-in filters. However, in some cases it may be necessary to install extra air filtration equipment at the source of supply air in order to meet the needs of a particular client or market segment; for example, an individual or family with severe allergies. Filtration devices that can be installed to remove particulate matter such as lint, dust and pollen from the air are discussed in Chapter 13.

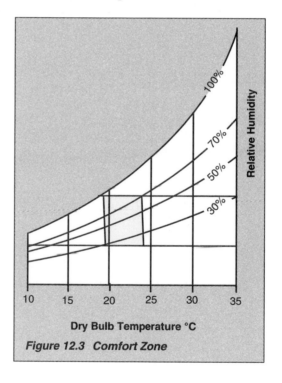

Figure 12.3   Comfort Zone

Chapter *13*

DISTRIBUTION SYSTEMS

As part of their space conditioning all houses need a means of efficiently distributing fresh air and exhausting stale air, and a means of distributing heating and/or cooling. There may also be requirements for filtration and/or humidity control of indoor air.

In most cases, space conditioning distribution systems can be combined. A properly designed forced-air system, for example, can be used to distribute heating, cooling and ventilation air. In other cases, separate distribution systems may be required to meet different space conditioning needs. For example, if individual room convectors are used to provide heating and/or cooling, a separate duct system will be required to distribute ventilation air.

Heating, ventilation and, in some areas, cooling systems are all regarded as essential elements of a home's mechanical system. However, there are other optional space conditioning systems which may be installed, depending on the needs of a particular market or the preferences of a particular client. Humidification and air filtration (and even cooling) systems fall into this category.

This chapter discusses the various distribution and control systems which can be used to meet the space conditioning requirements of a house.

The latter part of the chapter discusses the various types of equipment that can be used to humidify house air in winter, and to remove dust, pollens and other particulates from the air.

13.1 FORCED-AIR SYSTEMS

Only one distribution system can meet all the space conditioning requirements of a house: a forced-air system (see Figure 13.1). If properly designed, a forced-air distribution system can be used to distribute heating, cooling and ventilation air, and the air can be easily humidified, dehumidified or filtered. This is advantageous, since it is generally less costly to use a single duct system to distribute heating, ventilating and/or cooling air than it is to install separate distribution systems.

A typical central forced-air system is shown in Figure 13.2. In winter, the circulating air can help distribute heat from "point sources" such as solar gain or a wood stove through the house. This reduces temperature variations in the house and overall heating requirements. However, individual room or zone temperature control can be more complicated.

Thermostatically controlled dampers can be used, as can branch duct heaters with room thermostat controls (see Section 13.4).

Type of System	Space Conditioning That Can Be Provided					
	Heating	Zoning	Cooling Ability	Ventilation	Filtration	Humidification
Forced-air	X	X	Elaborate	X	X	X
Convection						
Electric	X		Easy			
Hydronic	X		Conventional	X		
Radiant	X		Easy			

Figure 13.1   Comparison of Space Conditioning Distribution Systems

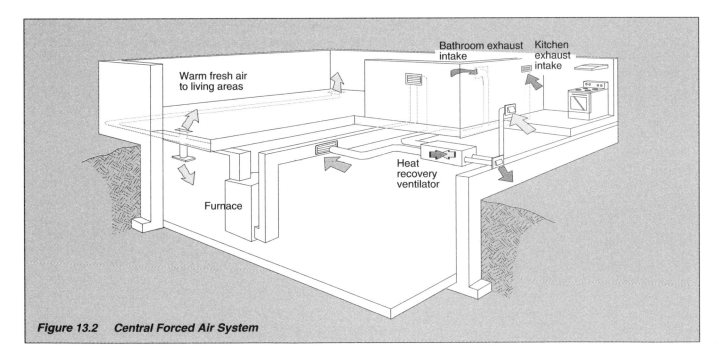

Figure 13.2   Central Forced Air System

13.1.1 Design Considerations

The basic sizing and design of the duct system should be based on the *Residential System Design Manual for Air Heating/ Cooling Systems* from the Heating, Refrigerating and Air-conditioning Institute of Canada (HRAI). The volume of heating air is generally much larger than that of ventilation air (see Chapter 16 for a detailed discussion of ventilation systems). Consequently, in a duct system designed for both heating and ventilation, the ducting is generally sized for heating but the outlet and return locations are chosen taking ventilation system design into consideration. The sizing of the branch ducts is based on a standard room heat loss and/or heat gain analysis.

If the ducting is used to supply ventilation air as well as heating and cooling air, the circulating fan must be operated continuously. This can add significantly to the energy budget of a home - energy consumption for a single speed fan operated continuously can be as high as 2500 kwh per year. In winter, heat generated by fans can contribute to space heat, but at an electrical cost, rather than a fossil fuel rate; during the rest of the year, it is an unwanted heat gain.

One solution is to install a two-speed fan — these run most of the time at the lower speed to provide circulation and ventilation, and only switch to high speed when called for by the occupant and/or the thermostat/furnace and/or the dehumidistat. Another solution is to install a furnace with a high efficiency circulating fan/motor (e.g. commutating type, with direct drive), which can use much less electricity. Such furnaces use as little as 1000 kwh per year, even with the fan on continuously. A good guide is to select a furnace that has the lowest electrical rating for the output that is required to heat the house. This information is usually included in the manufacturer's literature.

If the system must be designed to carry cooling air as well, larger ducts or higher speed operation may be needed because of the heavier air that cooling systems must carry, relative to heating systems. Most forced air heating systems introduce warm air into a room at a low level, since warm air naturally rises. Conversely, because cool air falls, it seems logical to provide cool air at a higher level. For distribution systems performing both functions, a good compromise is a floor system that can direct the cold air upwards with sufficient velocity to reach the ceiling (see Chapter 16).

Other forced-air system design considerations are summarized below.

- To maximize efficiency and to avoid a possible source of moisture damage, duct runs should not penetrate insulated and sealed walls or ceilings;

- Duct runs through unheated areas must be sealed with a suitable mastic at all joints and insulated to reduce heat loss. Consider insulating the warm air plenum and all major warm air ducts runs;

- To avoid condensation on ducts, cold air ducting for central air conditioning or outside air intakes must be insulated and provided with a sealed external vapour diffusion retarder;

- To minimize friction losses and to reduce fan power requirements, adequately sized ductwork should be laid out with smooth junctions, the lowest possible number of elbows and transitions, short runs and turning vanes for tight corners;

- Adequate cold air returns must be installed, including all the bedrooms;

- It is a good idea to seal all joints in all return and supply air ducts, especially those that will be hidden and inaccessible after construction is completed. Use a special duct sealing compound (mastic) which should be available from heating wholesalers. It is possible that certain metal foil duct tape may also work for this. Be sure that it will adhere well and will not peel off or become brittle with age;

- If the heating/cooling system is to be used to distribute ventilation air as well, the ventilation air will need to be tempered to ensure that the temperature of the air being discharged from the supply air outlets is above 18°C (65°F). Alternatively, use high inside wall supply outlets, to prevent cold drafts from causing occupant discomfort;

- If a mid-efficiency gas furnace is to be used with a combined heating/cooling/ventilating system, a calculation must be made to ensure that the mixed air (return air and cold outside air) temperature in the return air duct immediately upstream of the furnace will be above a certain temperature, as specified in CSA B326, or by the furnace manufacturer. Excessively cold air passing over the heat exchanger of this type of furnace may cause the combustion gases to condense. The resulting condensate is acidic and could prematurely corrode the furnace heat exchanger, which could, in turn, result in the release of toxic products of combustion into the house via the duct system;

- Any duct carrying air cooler than 13°C (55°F) must be insulated and covered with an air barrier to prevent condensation.

13.2   CONVECTION SYSTEMS

Convection systems are primarily used to distribute heat, usually from hydronic (hot water) central heating systems or from space heaters such as electric baseboard heaters or balanced flue gas-fired baseboards or wall-mounted heaters. Convection cooling systems are also available (see Chapter 15). If you install a convection heating and/or cooling system, you must still install a separate duct system for the distribution of ventilation air (see Chapter 16).

Convection heating systems rely on natural convection currents (warm air rises, cool air falls), sometimes with a fan assist, to distribute heat throughout a room (see Figure 13.3). Such units may be independent, as in room-by-room convector systems, or they may be connected, as in hydronic systems. In either case, air heated by convectors rises and is replaced by cooler air flowing in from the rest of the room, resulting in a convective loop with heat being distributed throughout the room.

A room-by-room heat loss analysis is required in order to size the individual room convectors. These in turn, can be controlled individually or in groups (by zones), or by using one central control. If more than one room is serviced by a central control, sizing of the convectors is especially critical. All spaces within a zone must have similar thermal responses and heating requirements. This usually leads to floor-by-floor control or north/south zoning.

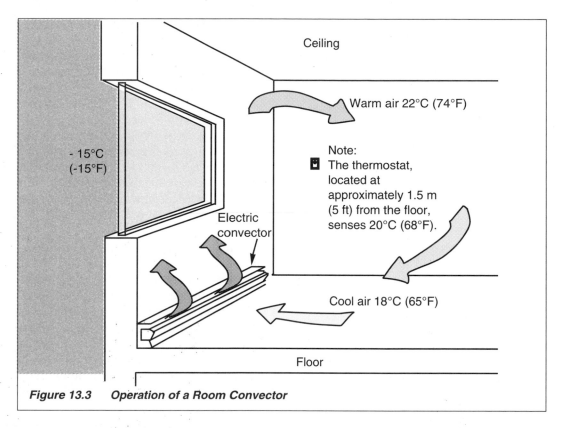

Figure 13.3   Operation of a Room Convector

Convectors have traditionally been placed at the floor level of perimeter walls, usually under sources of high heat loss such as windows. In locating convectors, try to take into account the anticipated placement of furniture and drapes, so that as the heat rises it can enter the space without major obstructions.

One drawback with convection systems is that a separate ventilation system must be installed. The lack of whole house air circulation limits options for cooling, (de)humidifying or filtering the air, as well as for making optimal use of point heat sources such as solar gains or heat from an advanced combustion wood fireplace or stove.

13.2.1   Room by Room Convector Systems

Individual room convectors make room temperature and zoning (see Section 13.4) particularly easy. The units can be housed in baseboards or recessed in floors, walls or cabinets, and occasionally come with small fans to improve circulation. At this point, electric baseboard heaters are the most common room convectors. While they have the advantage of no capital cost and no need to vent to the outdoors, they can be quite costly to operate. Direct vent gas-fired convectors, space heaters and efficient fireplaces, oil-fired space heaters, and advanced combustion wood-burning stoves and fireplaces (as described in Section 12.1.2) can be cost-effective alternatives. Some of these units, particularly the advanced wood stoves and fireplaces and some of the gas fireplaces can also function as radiant heaters, as described in Section 13.3.

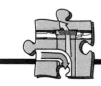

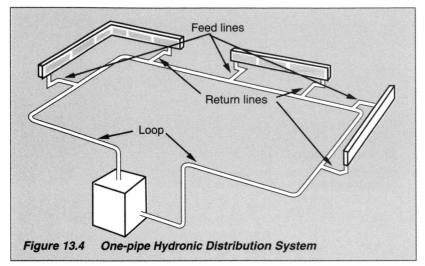

Figure 13.4   One-pipe Hydronic Distribution System

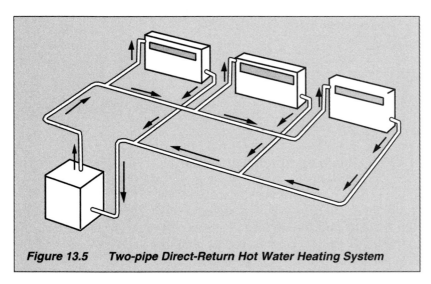

Figure 13.5   Two-pipe Direct-Return Hot Water Heating System

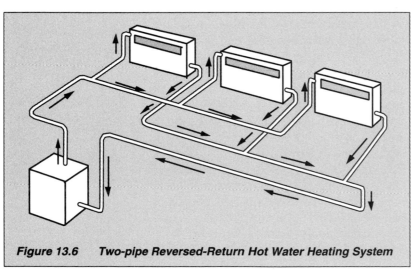

Figure 13.6   Two-pipe Reversed-Return Hot Water Heating System

13.2.2   Hydronic Distribution Systems

Hydronic distribution systems use water as the heat transfer medium. In such a system, the water is first heated in an oil-fired, gas-fired or electric boiler and then distributed throughout the house via a series of convectors (radiators). Some older systems relied on gravity and natural convection to move the water or steam, but modern systems all used forced circulation, with the design water temperature at 82°C (180°F) at the boiler exit.

There are three types of hydronic systems:

- In a *one-pipe system* (see Fig. 13.4), there is a single piping system to circulate the hot water. Each convector is connected to the main pipe by a branch feed line and return line. The system must be designed to ensure that the convector farthest from the boiler still provides enough heat. An advantage of one-pipe systems is that one or more single convectors in the system can be shut down without affecting the operation of the remaining units.

- In a *series-loop system*, the individual convectors are integrated right into the piping system with no branch pipes. The hot water flows sequentially through the entire system of piping and convectors. A disadvantage is that the convector closest to the boiler receives the hottest water. Those farther along in the system may receive water that is several degrees cooler; hence it becomes more and more difficult to heat the rooms furthest from the source.

- In a *two-pipe system*, the water returns from each convector to the boiler. The supply and return piping systems are separate, with branch pipes used to connect the convectors to both lines. In a direct-return system (see Fig. 13.5), the water returns to the boiler immediately after it passes through a convector. In a reverse-return configuration (see Fig. 13.6), the return water carries on through the system collecting the returns from all the convectors before it is returned to the boiler.

13.3 RADIANT SYSTEMS

Radiant systems can be a variety of different types. Usually, they use hydronic or electric resistance elements. Advanced combustion woodstoves and fireplaces and similar gas-fired appliances can also be effective radiant (as well as convective) systems, as discussed in Section 12.1.2.

In general, when the heat is first "turned on" with radiant systems, the occupants are warmed directly by the radiant heat; even though the overall room temperatures may be somewhat low. For this reason, radiant heat has a particular advantage for areas where the occupants may be susceptible to cold, such as bathrooms, or areas of high heat loss, such as entrance ways and rooms over floor slabs, or in major living areas, where the occupants spend a majority of their time.

Once the heat has been on for some time, other objects in the room also heat up and help warm the air to normal room temperature; at this point, the system works similarly to hydronic or electric baseboard systems.

- *Radiant space heaters* (see Figure 13.7) can be a very effective way of supplying heat and comfort to major living areas, and potentially lowering the overall heat requirements of the house. It is obvious that a wood stove is a good "black body" radiator. What is not so obvious is that the new advanced combustion wood-burning fireplaces, meeting CSA B-415 or EPA 1990 (see Section 12.1.2), are also effective radiant heaters.

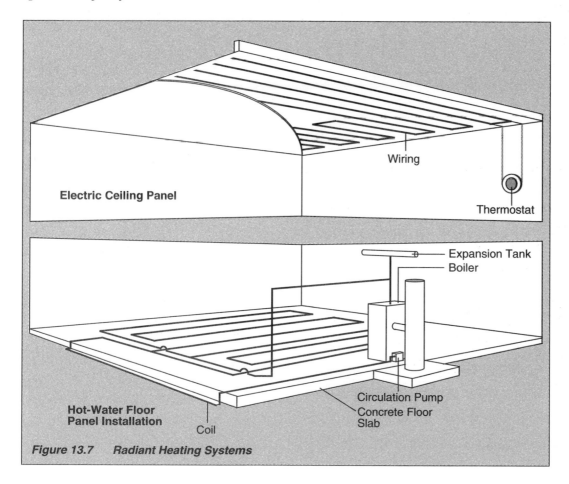

Electric Ceiling Panel

Wiring

Thermostat

Expansion Tank
Boiler

Hot-Water Floor Panel Installation

Coil

Circulation Pump
Concrete Floor Slab

Figure 13.7   Radiant Heating Systems

- The flame developed in these units is a good radiant heat source; the special ceramic glass which allows viewing of the flame is effectively transparent to infrared (IR) radiation, allowing nearly all the IR energy from the flame into the room. These new fireplaces are a means of cleanly and efficiently supplying a major portion of the home heating requirements, while using a renewable energy source. Some of the new efficient, direct-vent gas fireplaces, with ceramic glass windows, also function as effective radiant space heaters.

To some, the best way to use radiant heating systems is to heat ceramic tile or terrazzo floors that may otherwise feel cool, either by using hot water flowing through plastic pipes embedded in a cementatious material or by using electric resistance elements. This type of system must be installed with a thermostat connected to a sensor in the floor to keep the floor warm, but slightly below room temperature.

Floor coverings can dramatically reduce the benefits of radiant floor heating. Ceiling systems do not suffer this problem, although they can have higher heat losses.

- *Radiant Floor and Ceiling Systems* are another option. Radiant heating systems located in ceilings exposed to the outdoors lose some of their heat to the outside, even if the ceiling is well insulated. This can make this type of system less efficient than other designs.

In designing radiant floor or ceiling systems, you must take into account that only the area above or below the radiant heater will truly be heated. Consequently, to ensure comfortable conditions for a range of locations, you must consider the coverage as well as the capacity of the system.

Because radiant floor and ceiling systems are not common, they may require special design expertise and may be too expensive for a small energy demand. If you want to install such a radiant system, be sure to choose a knowledgeable designer.

In most cases, central heating systems or radiant space heaters/fireplaces will heat the space more effectively and economically than a radiant floor system, especially if electric resistance is used in the floor. There is no controlled-test evidence that radiant floor or ceiling systems are inherently more efficient, or even produce less heat stratification than conventional systems.

13.4   CONTROLS

Proper design of control systems is important. A well designed control system will help maintain comfortable conditions in the house in an energy efficient manner.

13.4.1   Thermostats

Temperature can be controlled by a central thermostat or by individual room thermostats.

- The location of a central thermostat is critical. It should be mounted on an interior wall, away from any local thermal effects which could "fool" it. Thermal effects include direct solar radiation, chimneys, plumbing vents, water pipes, fireplaces, wall ducts for heating and ventilating, or sources of drafts, including stairwells.

- Individual convectors allow room by room control. Built-in thermostats on baseboard convectors are inexpensive but generally provide very poor comfort control because they are located directly on the heat source. It's best to use them only in low occupancy areas. Wall-mounted thermostats are much more efficient in assuring overall room comfort. Electronic pulse thermostats which can sense temperature differences as small as ±0.5°C are available and may provide superior control.

Setback and programmable thermostats schedule indoor temperatures to automatically turn down when less heat is needed; at night, for example.

With some heating equipment and systems, controls can be used to smooth out temperature shifts by varying or "staging" the heat output according indoor heat gain/loss, and outdoor temperature changes. This helps to maintain even temperatures and improve comfort.

Outdoor sensors can be used to "kick in" additional heating capacity as the outdoor temperature drops. They can also start up the heating plant before the indoor temperature falls, in order to compensate for sudden changes in outdoor temperature.

If you install advanced control systems, be sure that the home owner understands how to operate and maintain them.

13.4.2   Zoning

In large, multi-level homes, or ones that are spread out, it can be difficult to achieve even temperatures in all rooms. This problem cannot be resolved simply by turning up the thermostat because some rooms will overheat. The solution is often to divide the house into two or more zones, each controlled separately. Separating the living areas from the bedrooms is one obvious approach. Bathrooms and entrance ways should be separately zoned to ensure comfort.

For forced air systems, you can use motorized dampers or valves to maintain the temperature balance in each area of the house. For hydronic heating, additional valves and pumps may be required. In Western Canada, some builders install two separate furnaces and distribution systems for different areas of the house.

13.4.3 Other Control Devices

Space conditioning equipment can include humidification and dehumidification devices to adjust the relative humidity of the conditioned air in the house. These devices are controlled by humidistats. Humidistats can also be used to control the operation of the mechanical ventilation equipment: once the relative humidity of the conditioned air rises above a certain preset level, the humidistat switches the ventilation system into high speed operation.

13.5 HUMIDIFICATION

A humidifier is a device used to add moisture to the air of a dwelling. The operation of a central humidifier can be controlled by a room or furnace-mounted humidistat. In leaky houses, humidifiers were a common addendum to many oil or gas furnaces. Today, with tighter housing, there is often not the same need to humidify in the winter.

Where systems are installed, the homeowner should be cautioned that they must be well maintained, or they can become sources of mould and fungi (e.g. legionella bacteria). Humidifiers on central furnaces have a tendency to leak over time, which could rust the furnace heat exchanger.

Central humidifiers use either an evaporative or a spray process to add humidity to the air, as described below.

13.5.1 Evaporative Humidifiers

There are three main types of evaporative humidifiers.

- In *by-pass (rotating drum)* humidifiers (see Figure 13.8), a drum wrapped with a water absorbent evaporator pad rotates, picking up water from a reservoir. As air passes through the pad it absorbs the moisture. This type of humidifier is installed on the return air plenum and connected to the supply air plenum with a flexible duct. Water is supplied through a 6 mm (1/4 in.) flexible tube connected to the nearest cold-water pipe.

- *Pan type* humidifiers consist of a water reservoir (pan), heating coils, and a fan with a motor. The heating coils, located in the pan, are warmed by low-pressure steam or forced hot water. Warm air is blown over the pan when the humidistat indicates that the air in the house is too dry. The fan shuts off when the relative humidity rises above the humidistat set point.

Housing

Rotating Drum

Water supply
Water reservoir

Figure 13.8   By-pass Type Humidifier

• *Stationary pan type* humidifiers consist of a pan and an evaporative pad. Warm air from the supply duct is passed over the pad by a blower or fan when the control system indicates that the relative humidity in the conditioned space is too low.

Minerals in the water can be a problem in evaporative humidifiers. Lime or other mineral deposits can build up on the water reservoir and on the evaporator pad. Some humidifiers are equipped with automatic flushing systems to clean out mineral deposits, while others must be cleaned manually by the home owner.

Moulds can also build up in the humidifiers if they are not cleaned regularly. These can be a source of poor indoor air quality and resultant health problems for the occupants.

13.5.2   Spray Type Humidifiers

These humidifiers work by breaking water up into fine mists. The mist is then converted to a vapour which is absorbed by the drier room air.

Spray type humidifiers consist of a chamber containing a spray nozzle system, a recirculating water pump, and a collection tank. As the air passes through the chamber, it comes into contact with the water spray from the nozzles. These humidifiers can be installed on either the supply or the return air plenum, with or without a controlling device.

Any minerals contained in water used in a spray type humidifier will remain in the mist. When the water in the mist evaporates, the minerals are left as a fine dust, which can be blown around the house through the distribution system. Some spray type humidifiers use filters to solve this problem.

13.6   AIR FILTRATION EQUIPMENT

Dust levels in a well-built home are generally low, and most heat recovery ventilators (HRV's) come equipped with basic air filters. However, in some cases, extra air filtration may be necessary to meet the needs of a particular client or market.

Electronic or charged media filters can be built into a forced air distribution system to remove dust, dirt, smoke, pollen and other fine particles from the air. Mechanical filters can also be installed over the supply air intake for the ventilation system. The homeowner should be made aware that air filters must be cleaned or changed regularly, according to the manufacturer's recommendations.

13.6.1   Electronic Filters

In electronic air filters (see Figure 13.9), air passing through the filter receives an intense positive electrical charge. Particles in the air, being positively charged, are then attracted to negatively charged plates or screens, where they are trapped. The best electronic air filters remove from 70 to 90 % of all the solid air contaminants.

Electronic filters are available as permanently mounted units which can be installed at the furnace, HRV or air conditioning unit, or in wall or ceiling return air grills. Stand-alone independent cabinet units are also available for cases where a permanently installed filter is impractical or where only selective air cleaning is needed.

If these units are not properly installed and maintained, they may be a source of undesirable ozone in the house environment.

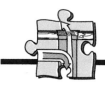

13.6.2 Charged Media Filters

Charged media filters work on a similar principle to electronic filters, but take advantage of electrostatic action to attract particles. The air passes through electrically-charged grids, creating a low-voltage electrostatic field. Dust particles in the air become polarized (i.e. one end of the particle is negative and one is positive) and are attracted to pads of glass fibre, cellulose or other material, in much the same way as iron filings are attracted to a magnet.

When these pads are filled, they must be removed and replaced with clean ones.

13.6.3 Controls

Built-in air filters can be wired to the blower of a forced-air heating or cooling system so that each time the blower is activated the filter will operate as well. The fan control on a room thermostat or combination thermostat/humidistat then controls the operation of the air filter, as well as the system blower.

13.6.4 Mechanical Filters

Mechanical filters can be used to remove pollen, dust and other particulates from the incoming air supply. The requirements for HRV installation state that if the HRV does not have a filter, then one must be installed in the duct or intake (see Appendix III).

Viscous impingement filters use a substance or medium coated with a sticky adhesive, oil or grease to trap particulates. The medium can be made of glass fibres, expanded metal, animal hair, nylon thread, or a combination of these materials. *Plastic foam* filters, made of open celled plastic foams, rely on surface irregularities to attract and hold particles. In general, mechanical filters are not as effective as electronic or charged media filters.

Standard furnace filters, usually made out of fibreglass, help to protect the circulating fan of a furnace from lint and accumulation and to remove some of the larger dust particles from the air. However, if any of the house occupants are sensitive to fine particulates due to allergies or asthma an electronic or charged media filter system is required.

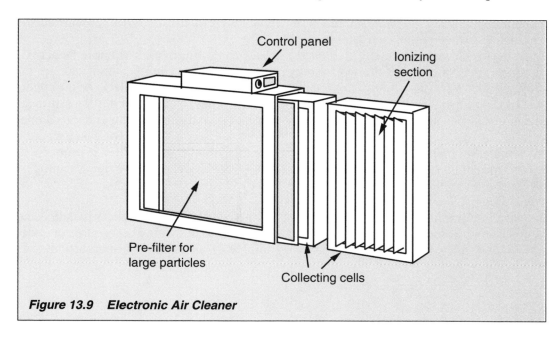

Figure 13.9   Electronic Air Cleaner

Chapter *14*

HEATING SYSTEMS

This chapter discusses the various types of heating systems suitable for energy efficient housing. The more energy efficient a home is, the smaller the heating system can to be. In houses built according to the techniques described in this manual, it may be possible to use simple, innovative, low cost heating options which are not practical in less efficient houses.

The following sections discuss the options available and the factors that you should consider in designing and installing the heating system.

14.1 DESIGNING THE HEATING SYSTEM

The heating equipment and distribution systems are usually selected at an early stage. To make sound decisions, you need to know:

- the design trade-offs between heating options and the construction of the house envelope
- the types of heating equipment currently on the market, and their relative efficiency, costs and performance, especially in terms of comfort
- cost, availability and local market preference for different fuels, including hook-up charges for low volume systems
- overall capital costs
- maintenance requirements, parts and servicing availability
- the implications of heat distribution options including customer preferences, space and locations available in the house, and the possible integration of the heat distribution and ventilation systems, and
- the possible integration of space and tap water heating.

Trade-offs can be made between heating and ventilating system efficiencies and overall envelope design. Using more efficient mechanical systems may allow you to use a less costly construction technique. When looking at heating, you should also look at the requirements for ventilation, domestic hot water and space cooling. It may be possible to combine or integrate certain components of the systems to reduce overall costs, while providing good performance and comfort levels. Be sure to choose an HVAC contractor with experience in installing the type of system you want.

Compare the *seasonal efficiency rating* (AFUE) for heating equipment: this is a better measure of relative annual fuel consumption than steady-state or peak combustion efficiency. For gas-fired equipment, check the Canadian Gas Association's *Directory of Certified Equipment.* For oil-fired equipment, many have U.S. AFUE figures. For wood-burning equipment, you should only install units meeting CSA B-415 or the U.S. EPA 1990 standard.

14.1.1 Energy Sources for Heating

The choice of heating fuel is influenced by:

- the personal preference of the home-owner
- the capital costs of associated heating systems
- the base and projected cost of the energy
- the efficiency in converting the fuel to useful heat
- regional or time of use fuel price differences, and
- availability.

Remote regions have some special considerations. In northern areas above the tree line, oil is often the best available energy source. Here, electricity is generated by oil-fired diesel generators and is often far too expensive for space heating. In northern areas below the tree line and in certain other regions where natural gas or fuel oil is not available, wood is often the main fuel for space heating. In other areas, it can be an effective supplementary heating source.

For homes with low space heating requirements, the difference in operating costs between alternate energy sources may appear to be relatively small. For example, a 20 per cent difference in energy costs may translate to only a $50-$60 saving on the annual fuel bill; a 50% difference would save $125-$150 per year.

The "lowest-cost" system may not always be the most satisfactory. Remember, too, that the costs of other energy uses, especially tap water heating, should also be considered. In some cases, choosing a slightly higher cost fuel for space heating can save as much as 45% on water heating.

Moreover, many utilities have minimum prices or service charges, such as monthly billings, which can significantly affect the cost per unit of energy at low usage rates. In some cases, a utility may even impose a premium to hook up a system for a house with low projected fuel consumption.

To compare between different energy sources, it is important to know the unit energy costs of each source. Figure 14.1 presents the unit energy values for most common sources in Canada. To determine the relative operating costs, multiply the unit energy value by the seasonal efficiency of the appliance, the unit price of the energy source and the projected energy demands of the house.

Energy Source	Energy Content
Fuel Oil	38.2 MJ/Litre
Natural Gas	37.5 MJ/m³
Propane	25.3 MJ/Litre
Electricity	3.6 MJ/Kwh
Hardwood (air dried)	25000 MJ/cord
Softwood (air dried)	18700 MJ/cord
Wood Pellets	19800 MJ/tonne

Figure 14.1   Energy Content of Various Energy Sources

14.2   GAS-FIRED SPACE HEATING APPLIANCES

In many regions, natural gas is an economical energy source and hence is very popular. In more remote locations, propane serves as an alternative, yet more expensive option.

Until recently, most of the gas furnaces sold were of the conventional type, with a continuous pilot light, naturally aspirating burner, "clamshell" heat exchanger and natural draft exhaust. The seasonal efficiency of such equipment was on the order of 60-65%. Some units also had a flue damper, to close off the chimney when the furnace was not operating; their real seasonal efficiencies were just slightly better.

Today, only mid or high-efficiency gas-fired furnaces and boilers are produced. The following are some of the efficient appliances available which can operate on either natural gas or propane.

14.2.1   Mid-Efficiency Gas-Fired Furnaces

This type includes both induced and forced draft furnaces, as described in 14.2.1.1 and 14.2.1.2, respectively. Both types are designed to eliminate the need for dilution air (i.e. no draft hood), are usually vented up a chimney or sidewall vented, with seasonal efficiencies in the range of 78% to ± 87%.

Units should be carefully sized to the house heating requirements. An oversized unit will have a lower realized efficiency, and might result in comfort problems. Ironic as it sounds, mid-efficiency furnaces with ratings higher than 82% should be avoided. They often have, or are susceptible to, problems with condensation and corrosion in either the furnace heat exchanger or the venting system. If you want to achieve higher efficiencies, go to a condensing system, as described in Section 14.2.2.

Mid-efficiency furnaces fired with propane will tend to be 1-2% higher efficiency than the same units fired with natural gas.

Natural draft mid efficiency gas units have both lower efficiencies and potential draft problems, so they are not recommended for the type of house described in this manual.

Mid-Efficiency Gas: Induced Draft Furnaces/Boilers

These units use a fan in the flue, downstream of the furnace proper, to induce or pull the combustion products through the furnace and propel them out the vent. The burner itself is usually a conventional, naturally aspirating type but it does not have a continuous pilot light. There is no draft hood (see Figure 14.2). Some of these furnaces can be sidewall-vented, although the majority use a chimney.

Mid-Efficiency Gas: Forced Draft Furnaces/Boilers

These units have a fan or blower upstream of the furnace heat exchanger. It is usually a blower which supplies combustion air for the burner (power burner), and acts to push the combustion products through the furnace and out the vent. There is no draft hood and the burner typically has an intermittent ignition device. This is more commonly found on boilers for hot water heating systems than on warm air furnaces. Depending on the particular appliance, there may be options for vertical or sidewall venting.

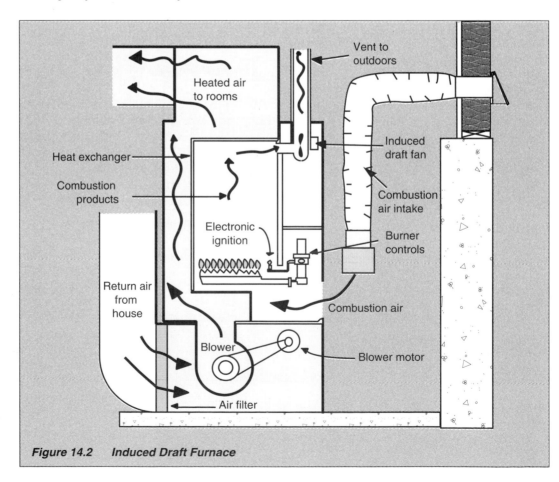

Figure 14.2   Induced Draft Furnace

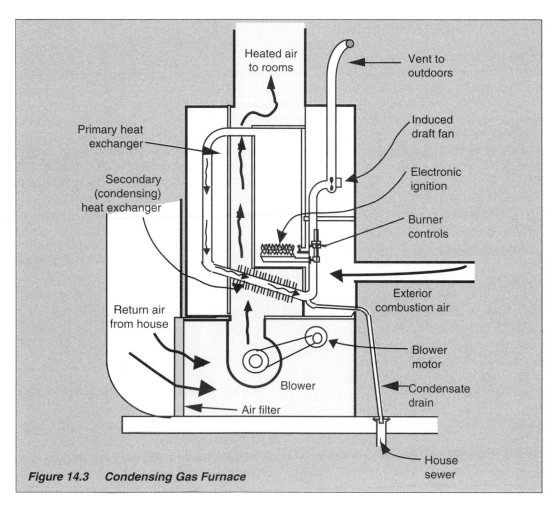

Figure 14.3   Condensing Gas Furnace

Labels in figure:
- Heated air to rooms
- Vent to outdoors
- Primary heat exchanger
- Induced draft fan
- Electronic ignition
- Secondary (condensing) heat exchanger
- Burner controls
- Exterior combustion air
- Return air from house
- Blower motor
- Condensate drain
- Blower
- Air filter
- House sewer

14.2.2   High Efficiency Condensing Gas Furnace

Condensing furnaces, also known as high efficiency units, have seasonal efficiencies in the range of 92% to 96%. While most condensing furnaces use a conventional aspirating burner and induced draft fan (see Figure 14.3), alternative combustion designs exist, such as pulse combustion, infrared ceramic and power burners.

High quality heat exchanger in these furnaces cool the combustion products below their dewpoint, condensing out water vapour in the flue gas and regaining the latent heat. Usually the temperature of the exhaust is so low (35°C to 50°C) that a plastic pipe is used to carry the gases to a sidewall vent. There is no need for a chimney.

The condensate is slightly acidic requiring corrosion resistant materials — most are made out of high grade stainless steel. The exhaust piping must also resist corrosion. The condensate itself is collected, neutralized and usually discharged directly to the drain.

Condensing furnaces work well with a shorter cycle length: more water is condensed, due to cold furnace walls, and higher efficiency. Thus a certain amount of oversizing is allowed with a condensing furnace. Indeed, it might actually result in slightly higher seasonal efficiencies.

Some of the units utilize a closed combustion system, with air drawn directly from outside to the burner and exhausted through a different pipe.

14.2.3 Gas Fired Boilers

As described above for warm air furnaces, there are two basic levels of equipment for gas-fired boilers; mid and high-efficiency, the latter being "condensing" type. Either one can be used to supply the heat directly to an hydronic heating system or indirectly through a fan coil to a warm air heating system.

One problem with high-efficiency condensing boilers is that they often do not achieve their high efficiency potential in practice, because the return water temperature is above the dew point of the flue gasses — there is no temperature differential to drive the condensation process. Coupling an upstream, in-line storage water heater with the condensing boiler, as shown in Figure 14.4, can regain some of this high efficiency potential, while generating hot tap water at a high efficiency level, as well. More details of this are presented in Chapter 18.

14.2.4 Gas Fireplaces and Space Heaters

As described in Sections 13.2 and 13.3, radiant/convective heaters can be an effective way of heating a local area of the house. Some gas-fired appliances develop an attractive visual flame and have a special pyro-ceramic glass door to allow viewing, while transmitting radiant heat to the room. With gas fireplaces, convection fans extract more heat from the appliance and pump it into the room. Controls for these units can either be a wall-mounted thermostat or merely a hand-held controller, similar to those now used for VCR's. These fireplaces often have direct vent or induced draft fan exhaust. In particular, there are new, efficient gas fireplaces which can satisfy consumers aesthetic need to watch a fire, while being a clean, efficient source of energy for the house.

The Canadian Gas Association is developing an efficiency standard for gas fireplaces (CGA P4-1993). It should also help builders in their choice of product.

Other gas-fired convective heaters, some of which resemble baseboards, (see Section 13.2) which can be side-wall vented.

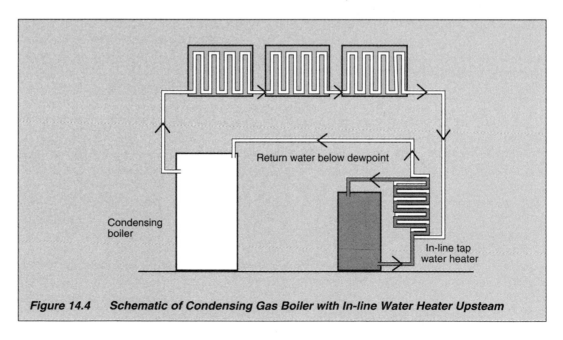

Figure 14.4   Schematic of Condensing Gas Boiler with In-line Water Heater Upsteam

14.2.5　Natural Gas vs Propane

In general, the same technologies and comments apply to propane as to natural gas, with slight differences in the efficiencies. Propane has a lower hydrogen level than natural gas. This means that less energy (about 3%) is tied up in the form of latent heat with propane systems than with natural gas. In other words, conventional and mid-efficiency propane furnaces can be expected to be slightly more efficient (1-3%) than natural gas units. On the other hand, the lower hydrogen content makes it more difficult to condense the combustion products, so that condensing furnaces will be about 1-3% less efficient on propane than on natural gas.

Another factor to consider is energy cost. Propane and natural gas are not usually sold on the same basis. Natural gas typically has 37.5 MJ/m^3, while propane has 25.3 MJ/l. On the average in Canada, natural gas and heating oil are from 35-60% cheaper on an energy basis than propane.

14.3　OIL-FIRED SPACE HEATING APPLIANCES

Oil is still the fuel of choice in many regions, particularly those without access to natural gas. It has the advantage of being a compact, storable fuel which is much less expensive than electricity or propane. In comparison to gas, oil has about half the hydrogen content, resulting in a much lower latent heat loss for oil. This accounts for the fact that comparable, non-condensing furnaces and boilers usually have a lower efficiency when fired with natural gas than with oil. Until the mid-1970's, oil furnaces and boilers had a cast iron head-type burner. The resulting seasonal efficiency of such equipment was generally in the 60% range. The replacement of this burner with a flame retention head burner, has improved efficiency to a seasonal level of 72-75%.

14.3.1　Flame Retention Head Oil Burner

Today all new oil furnaces are fired with some type of flame retention head burner. Depending on the burner and furnace technology the AFUE seasonal efficiency ranges from 78% to over 90%.

Some new *advanced flame retention head burners* operate at a high static pressure, enabling them to run at lower excess air levels (with commensurate efficiency gains) while overcoming pressure fluctuations generated at the vent termination. The pressure drop across the burner head also precludes the loss of heated house air through the burner and furnace and out the vent during the off-cycle.

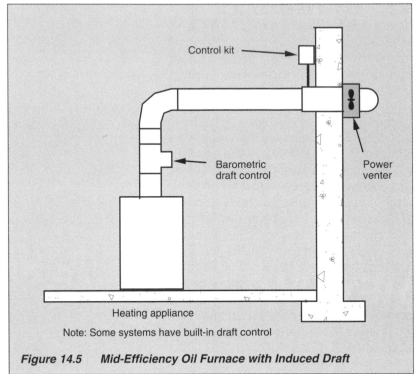

Control kit

Barometric
draft control

Power
venter

Heating appliance

Note: Some systems have built-in draft control

Figure 14.5   Mid-Efficiency Oil Furnace with Induced Draft

14.3.2   Mid-Efficiency Oil Furnace (Forced or Induced Draft)

The mid-efficiency oil furnace effectively eliminates the barometric damper, with its large air requirement. It may have an induced draft fan, located downstream of the furnace proper (often at the side wall of the house), which pulls the gases through the furnace and propels them up the stack or out the side wall of the house (see Figure 14.5).

An alternative is forced draft, where a high static pressure drop burner is coupled with a properly baffled furnace, allowing the burner to withstand any pressure fluctuations transmitted from the top of the stack (see Figure 14.6).

Knowing the exact characteristics of the flue gas allows closer tolerance of the venting system; more heat can be taken out of the gases without fear of condensation and corrosion. It is possible to eliminate the chimney altogether, exhausting the combustion gases out the side wall of the house. The seasonal efficiency of a mid-efficiency oil-fired type furnace can reach nearly 90%, without condensing.

Compared to conventional flame retention head-equipped furnaces, real fuel savings would be 5-10%. Additional benefits are greater customer satisfaction, reduced reliance on natural draft, much lower total air requirements and a safety shut-off in the event of flue blockage or reversal.

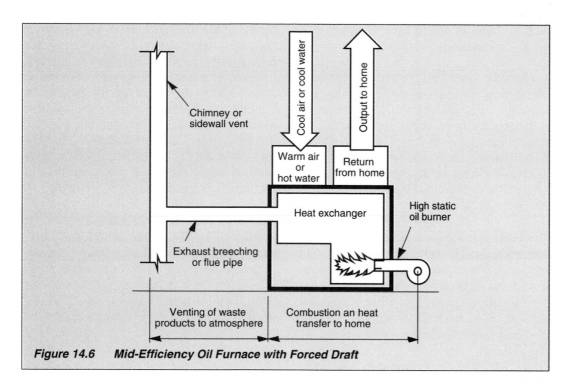

Figure 14.6   Mid-Efficiency Oil Furnace with Forced Draft

14.3.3   Condensing Oil Furnace

A further possibility is a condensing oil furnace, analogous to those discussed previously for natural gas. This type of furnace is designed to recover some of the latent heat loss by condensing water vapour from the flue gases. This takes place in an additional heat exchange section made out of stainless steel, or plastic with a water spray. The temperature of the furnace exhaust gases is lowered below th dewpoint, thus regaining the latent heat.

However, with oil producing much less water vapour and having a lower dewpoint than natural gas, the potential for efficiency improvements by condensing the flue gas is much lower for oil; you have to work harder to condense less. Also, high sulphur levels make the condensate very corrosive, so that any condensing heat exchanger must be extremely corrosion resistant. Oil combustion produces soot which can concentrate in the acidic condensate, producing "acid smut" at certain points on the heat exchange surface and make things even more difficult.

Condensing oil furnaces range from somewhat less efficient to as much as 8% more efficient than the mid-efficiency types described previously.

Some condensing oil systems use two vents, maintaining a conventional chimney, along with a plastic pipe out the side wall of the house, both open at the same time. With this type of system, there is a strong possibility of the flue gases bypassing the water spray condensing system and going straight up the chimney. The barometric damper is also retained, so that the additional dilution air increases the overall heat loss, while lowering the dewpoint even further, making it even harder to condense the flue gases.

At this time, the higher costs and potential corrosion of condensing oil units make the mid-efficiency oil furnace appear to be the preferred choice.

14.3.4 Use of Oil in Low Energy Housing

A significant problem appears to exist with oil heating systems in new, low energy housing. Because of the limitations in burner technology, firing rates cannot be reduced much below 0.5 US gph (70 kBtu/hr) and minimum furnace outputs of about 60 kBtu/h. Well-insulated new homes can have a design heat load between one-third and two-thirds of this level.

However, an effective solution exists. Combining the space and water heating in an integrated system may allow efficient operation of the furnace/boiler without short cycling and the attendant losses in efficient and comfort. In such a fashion, oil can indeed be an ideal energy source for new housing.

Figure 14.7 shows a basic system, with the energy generator, usually a low-mass boiler, coupled to a well-insulated storage tank by an efficient water-to-water heat exchanger. The better systems also come with intelligent microprocessor controls and a reversible pumping system. When the thermostat calls for heat, the boiler runs normally.

When the thermostat is satisfied, the burner continues to run instead of shutting off, but the hot water it generates is passed through the heat exchanger, sending heat to the storage tank instead of the house. As the thermostat requires heat, the process reverses, and heat is drawn out of the storage tank, across the heat exchanger, and distributed around the house, without the boiler actually operating. If hot tap water is required, it is taken directly out of the storage tank.

Both the overall operational time and the number of cycles of the boiler would be much less than a conventional system. As the boiler operates more efficiently in its heating mode, the tap water heating efficiency is improved dramatically. This can eliminate the need for a separate electric or oil-fired water heater.

Tap water heating has often been done by a tankless coil in an oil-fired hot water boiler. This system can be very inefficient, especially in summer. Inefficient short-cycling is common. Off-cycle stack losses are high as is the potential for casing losses. The system described in the previous paragraph eliminates these problems, while improving efficiencies.

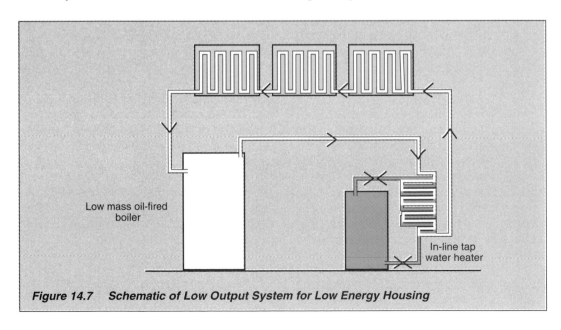

Low mass oil-fired boiler

In-line tap water heater

Figure 14.7   Schematic of Low Output System for Low Energy Housing

14.4 *ELECTRIC SPACE HEATING APPLIANCES*

Electric resistance heating systems with either central furnaces/boilers or electric baseboards are used by many builders. They are relatively easy to size and control and do not require either combustion air for operation or a venting system for combustion products. As a rule of thumb, all electric resistance heating systems are assumed to be 100% efficient.

However, many provinces are now strongly discouraging electric heating as part of demand side management initiatives. Others allow it in conjunction with a fossil fuel-fired system (e.g. bi-energie in Quebec). In Ontario, builders wishing to use electric heat will have to increase insulation levels to near the R-2000 level, use heat recovery ventilators and take other energy conserving measures if electricity is used for space heating. (Note: houses with earth energy systems, as discussed in Section 13.4.3.2 will not have to meet all these extra requirements.) In most areas, the operating cost of electric resistance heating will be significantly higher than for fuel-fired heating, rapidly outstripping any first-cost savings of the resistance heating system.

14.4.1 Electric Furnaces and Boilers

Electric furnaces use a forced air distribution system and a centrally located thermostat. An electric furnace is very simple, consisting of a number of resistance coils and a circulating fan, the coils capable of being operated with staging controls to limit swings in temperature. In a small house with a very low heat load, it may even be possible to use a simple in-duct heater and fan to provide heating and air circulation. Electric boilers, with resistance coils immersed in water, are available for hydronically heated homes.

14.4.2 Baseboard Heaters

Electric baseboards and other room-by-room systems allow each room to be individually thermostatically controlled, either with a low voltage thermostat or a CSA-certified line voltage wall-mounted thermostat. This type of system has been popular because of its very low initial costs and ease of installation.

Available electric heating elements include baseboard convection units, and radiant heating panels.

In general, it is very difficult to heat a room effectively with such units; often the baseboards are located at the outside walls, under high heat loss areas such as windows. Moving the heat to the rest of the area may pose problems.

If you must use this type of system, consider the surface temperature of the heating element. Lower temperature units may eliminate the problem of dust charring and may have some comfort and air quality benefits.

The advantages of individual room control and cheap initial costs have to be weighed against the high cost of operation for the homeowner. In addition, a separate distribution system must be installed for ventilation air at a significant cost. In general, electric baseboards are not seen as a desirable primary source of heating at this time.

14.4.3   Heat Pumps

Electric heat pumps (see Figure 14.8) can provide both space heating and air conditioning. The earth energy type described in Section 14.4.3.2 are capable of supplying hot water as well.

A heat pump usually has a heat exchange coil containing a refrigerant (usually freon, a CFC, at this time) which extracts heat. A significant amount of electric energy is required to "drive" the heat pump but, under many conditions, the heat it is capable of supplying to the house is more than would be produced by an electric resistance furnace. The ratio of heat supplied (to the house) to the electrical energy expended is defined as the coefficient of performance (COP) - the higher the COP, the more efficient the heat pump. The COP drops as the difference between the heat source (supply) temperature and the use (house) temperature increases.   When the COP reaches a value of 1, the performance is the same as an electric resistance furnace.

Since the electric energy used by the heat pump is effectively constant, a dropping COP means a decreasing output. That is to say, the heat pump can supply a lower amount of heat per unit time. This often occurs in the opposite direction to need. The colder it is outside, the more heat the house requires, yet less and less heat is available.   This problem is particularly applicable to air source heat pumps, as described in the next section (14.4.3.1).

Air Source Heat Pumps

With air source heat pumps, the heat is taken from the cold outside air. However, the colder it is outside, the less heat is available from the heat pump with its decreasing COP. Indeed, the best efficiency of most air source heat pumps is at around 7°C, a level at which an energy-efficient house really needs very little heat.

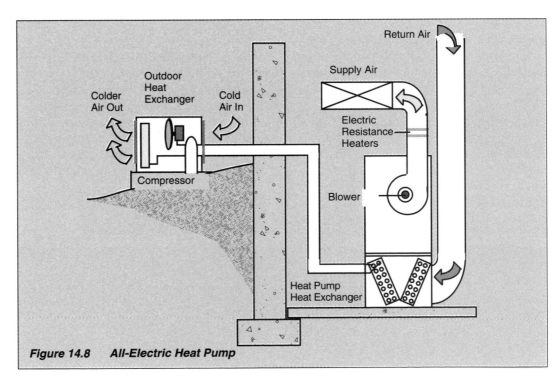

Figure 14.8   *All-Electric Heat Pump*

All-electric, air source heat pumps have to rely on back-up electric resistance heat to provide supplemental heat when the COP is too low. There are also air-source heat pumps coupled to fossil-fuel-fired furnaces. With such units, the heat pump is shut off completely when the outside temperature falls below a certain point, and all the house heat requirements are supplied by the fossil-fuel-fired furnace. With the latter system, problems are sometimes encountered if the choice of heat pump or furnace, instead of being controlled by the outside temperature, is driven by a two-stage thermostat. This can result in short cycling of the fossil-fuel furnace: an inefficient and potentially corrosive mode of operation.

Earth Energy Heat Pumps

Earth energy systems, also known as ground or water source heat pump systems, use the earth or groundwater as a source of heat in the winter and as a heat sink in the summer (for air conditioning). Since the ground or groundwater below the frost line remains relatively constant in temperature throughout the year (well above outdoor temperatures in winter and below them in summer), it provides a constant heat source for a heat pump to heat both the house and the hot water. The initial cost of an earth energy system is higher (sometimes much higher) than other systems, but the efficiency in terms of COP is very high (3-3.5), so that the operating costs are much lower than for other electric systems. In some areas, earth energy systems outsell air source heat pumps.

14.5   WOOD-FIRED SPACE HEATING APPLIANCES

New wood-fired space heating appliances are available in a variety of forms: stoves, furnaces, boilers, and fireplaces for both heating and recreational uses. For heating, wood fuel is only practical where there is an ample supply of local, low cost wood.

Wood heating requires significant user commitment and must be evaluated from a marketing perspective. For some consumers, wood heat is the fuel of choice, providing self sufficiency, security of supply, greater comfort, utilization of a renewable resource, and a low cost source of energy. Consideration should also be given, however, to the harsher realities of heating with wood.

- Wood fuel is heavy, messy, and requires considerable storage space when brought into main living areas. The introduction of pellet fuels may alleviate this problem.

- Wood brought indoors to dry may increase humidity levels and contribute to excessive condensation. It may also contain destructive insects.

- For consistent heat, the fire must be fuelled at regular intervals.

- A back-up heating system should be provided to ensure the home's temperature doesn't fall too low during extended absences or holidays.

- The user must become knowledgeable about the heating system's operation and maintenance requirements to optimize efficiencies and avoid hazardous situations.

14.5.1 Conventional Fireplaces

Conventional fireplaces are between -5% and +5% efficient. In other words, they supply little if any energy to the house, particularly when it is cold outside. One of the main reasons for the ridiculously low efficiency of a conventional fireplace is its large air requirements (about 1.5 air changes per hour), which suck massive quantities of heated air out of the house. In some conditions, smouldering fireplaces can produce hazardous pollutants indoors. Furthermore, depressurization caused by the fireplace can also affect the performance of other combustion appliances, causing them to spill combustion products into the house. Finally, conventional fireplaces emit large quantities of outdoor air pollutants in the form of suspended particulates. Devices such as glass doors, "heatilator" type heat exchangers and even outside air supply improve the efficiency marginally (up to a maximum of perhaps 20%) but have only a limited effect on the indoor and outdoor pollution problems.

The only way to improve efficiency and reduce these emissions is to change the combustion process, as with the advanced combustion fireplace as described in Section 14.5.4.

14.5.2 Space Heating Wood Stoves

Wood can be a useful energy source in a well-designed airtight wood stove, properly located in a major living area and vented directly into its own inside chimney, or into the fireplace chimney if that has been fitted with a stainless steel liner.

Unlike other fuels, wood does not burn away evenly. This means a number of waste products (CO, hydrocarbons, particulates and creosote) come off the wood as it burns. The "slower" the burn, the more products of incomplete combustion are released. In conventional airtight stoves, a large amount of these incomplete combustion products escaped the burning process to be deposited in the chimney as creosote and/or to be emitted to the environment as serious air pollutants.

14.5.3 Advanced Combustion Wood Stoves

In order to ensure clean, efficient combustion in the firing range required, major changes to the combustion design of conventional wood stoves were required. New stove designs are now being developed which give better combustion and have lower heat outputs, to yield a more useful range of operation. There are a number of new designs, employing either advanced combustion techniques or catalysts to reduce the amount of incomplete combustion products and increase efficiency.

Advanced combustion wood stoves now in the marketplace (see Figure 14.9) show an 80% reduction in emissions of incomplete combustion products with a 10-20% gain in efficiency, relative to stoves of five years ago.

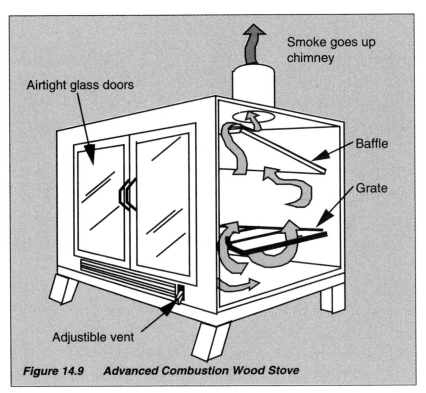

Smoke goes up chimney

Airtight glass doors

Baffle

Grate

Adjustable vent

Figure 14.9   *Advanced Combustion Wood Stove*

Such appliances can be an effective complement to conventional heating systems in many regions of the country; they offer the potential to displace 60-70% of the fossil fuel used for central heating in these regions, with a similar reduction in overall CO_2 emissions. They are also ideally suited for use in electrically heated homes, easily displacing 70% of the electricity used for space heating. Space heaters fired by pelletized agricultural or other biomass wastes offer another alternative to this application.

14.5.4 Advanced Combustion Wood Fireplace

Similar advanced wood combustion designs are now being utilized by Canadian manufacturers to produce what can be called an advanced combustion wood-burning fireplace. Such a unit has air tight doors; a hot-air-swept, pyro-ceramic glass window to allow clear viewing of the flame and good infra-red heat transmission; an insulated outer casing; good heat exchange and an effective circulating fan to supply heat to the house (see Figure 14.10). Most importantly, as with the stoves described in the previous paragraph, the emissions of incomplete combustion products are reduced to one tenth of those from a conventional fireplace. They can supply energy to the house at about 70% efficiency. In order to ensure this performance, you should get a unit meeting the emissions criteria of either *EPA 1990 or CSA B-415*. Indeed, in the R-2000 housing program, only wood-burning fireplaces meeting these criteria are allowed. An alternative is to install one of the new, efficient gas fireplaces, as described in Section 14.2.4.

14.5.5 Central Wood-Fired Systems

Central chunk wood-burning furnaces and boilers may appear to have attractive features: the fuel is restricted to the basement, the heat is distributed through the house via a central system and the units are controlled by a thermostat which regulates the heat output by cycling the furnace. However, contrary to the practice of on-off cycling with oil or gas furnaces, woodburning furnaces/boilers cycle between high and low fire. This produces large amounts of incomplete combustion products on each cycle change. In spite of the more difficult burning regime, the combustion technology is much more primitive than in the advanced combustion stoves and fireplaces described above. Pollutant and creosote production is high and the efficiencies are low, often in the 40% range. As the furnace gives no visual benefit to the occupant by viewing the flame, supplying and burning wood for one of these inefficient central heating furnaces/boilers can become an unrewarding daily chore.

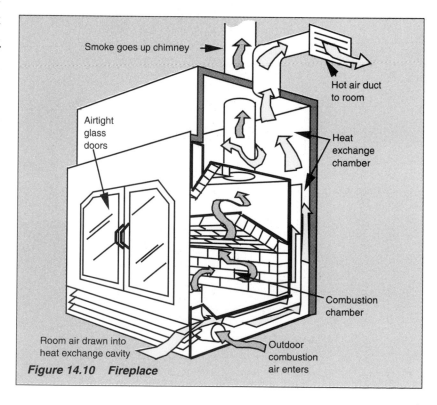

Smoke goes up chimney

Hot air duct to room

Airtight glass doors

Heat exchange chamber

Combustion chamber

Room air drawn into heat exchange cavity

Outdoor combustion air enters

Figure 14.10   Fireplace

Until such time as there are chunk wood-fired furnaces or boilers tested and approved to CSA B-415, it is recommended the advanced combustion wood space heaters and fireplaces (as described in the previous sections), be considered.

14.5.6 Wood Pellet-Fired Appliances

An alternative to burning wood logs is to use pellet burning space heaters, fireplaces and furnaces, which can be even more efficient and cleaner burning. They offer clean burning, automated fuel feed from a storage hopper and the potential to do without a chimney by side-walling the combustion exhaust. Their main drawback is the price of the fuel, which can be significant in many areas.

Chapter **15**

COOLING SYSTEMS

In well-insulated houses, cooling requirements are generally lower than in a conventionally constructed building. Nevertheless, in many parts of the country, cooling systems will be demanded by buyers.

Central air conditioning can be an attractive marketing feature in many areas and has the added benefit of dehumidifying the air. Most residential air conditioning equipment uses a compressor cycle: a refrigerant is alternately compressed and expanded. The cycle works on the same principle as a heat pump, only in reverse. Gas-fired air conditioners are at a developmental stage.

This chapter looks at the different options for central air conditioning.

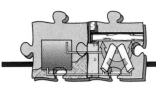

15.1 COOLING EQUIPMENT

Most air conditioning equipment is *compression equipment* which alternately compresses and expands a refrigerant (see Figure 15.1). This equipment works in the same manner as an electric refrigerator.

- Liquid refrigerant is forced under high pressure to an *expansion device*. As the high pressure liquid passes through the expansion device into the *evaporator*, it absorbs heat from the warm air flowing across the outside of the evaporator, expands into a larger volume, and vaporizes.

- The vaporized refrigerant is then pumped to the *compressor* where it is compressed and forced into the *condenser*. Cooled by air or water passing over the condenser coils, the hot vapour liquefies. It is then ready to start the cycle once more. Gas-fired air conditioners are still in the development stages.

In *absorption-cycle equipment*, heat (usually from a gas flame) drives the cycle: a refrigerant (usually ammonia) is boiled out of an absorbent (usually water) and is expanded in an evaporator coil where it absorbs heat from the surrounding air or water. The refrigerant then passes through a water-cooled condenser and is reabsorbed to start the cycle again. This is the same process that is used in a propane refrigerator.

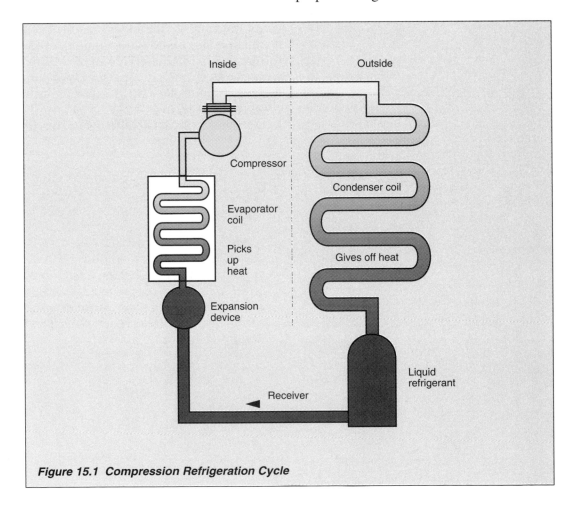

Figure 15.1 Compression Refrigeration Cycle

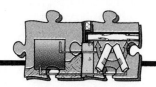

15.2   SYSTEM DESIGN

In designing a cooling system, it's important to choose efficient equipment. Manufacturers rate equipment in terms of its Seasonal Energy Efficiency Ratio (SEER). An efficient unit will have an SEER of 10 or greater. In some provinces, equipment with efficiencies lower than 10 cannot be sold.

Most central air conditioning systems circulate cooled air through the house distribution system. However, systems adaptable to hydronic distribution systems are also available.

15.2.1   Ducted Air Systems

Most air conditioning systems use the heating distribution system to circulate cooled air during the warm months. While more expensive, it is possible to provide a separate duct system for cooling.

The *design of the duct system* and the *position of air supply vents* is different for combined versus separate systems.

Ducted air systems circulate cooled air through the house. They can also be equipped with air cleaning devices to provide air filtration.

The compression/condenser unit is located outdoors, but the expansion/evaporator unit may be located indoors. The cooling equipment can be linked to the heating system or installed independently.

• In one option, furnace-mounted cooling coils are connected to a remote outdoor condenser unit. The evaporator coils are mounted in the plenum and the system uses the furnace blower to provide air circulation. The refrigerant is carried in piping that connects the indoor coils to the outdoor condenser unit (see Figure 15.2).

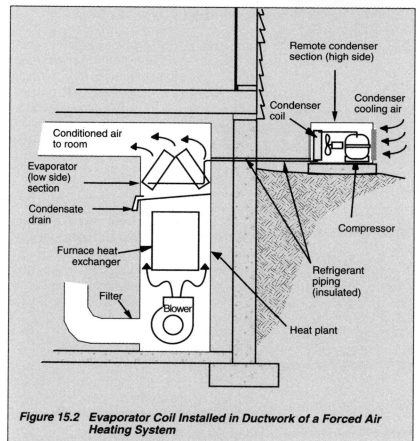

Figure 15.2   Evaporator Coil Installed in Ductwork of a Forced Air Heating System

• In a self-contained system, the entire cooling system, excluding the duct system, is contained in one enclosure located outside the house. Openings must be provided in the building envelope to allow the insulated supply and return air ducts to be connected to the indoor duct system. These openings should be sealed to the air barrier in the same manner as the electrical service conduit or water supply pipe (see Chapters 8 and 9).

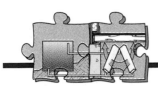

Combined Duct Systems

These are less expensive, easier to install, and take up less space than two separate systems. However, the homeowner may have to adjust duct dampers and blower speeds seasonally to ensure that each room is conditioned appropriately. Cooling systems tend to require a greater volume of air flow than heating systems, so larger ducts may be necessary.

Since warm air rises, most heating systems introduce warm air into a room at a low level. However, cool air falls; ideally cooled air should be provided at a high level. For combined systems, the best compromise is a floor unit that directs the air flow upward with sufficient velocity to reach the ceiling.

Separate Systems

Where the cooling system has separate ducts and return air intakes, they should be located in the ceiling or high on the walls. This type of system is simpler to design, and since the outlets are high, there is less likelihood that the occupants of the house will experience uncomfortable drafts. However, separate duct systems cost more, and require more space. If attic- or roof-mounted equipment is used, it may be difficult to service.

15.2.2   Chilled Water Systems

Chilled water systems can be used separately or with the distribution system of a hydronic heating system to provide summer cooling. Water, rather than air, is passed over the evaporator coil where it is cooled and then pumped through the system.

Typically, the piping system for cooling is totally independent of the heating system. This allows both heating and cooling systems to be designed for maximum efficiency.

Each convector (see Figure 15.3) consists of a coil through which chilled water is circulated, a fan to blow air over the coil, and a filter which removes particulates.

A separate convector is generally located under a window or recessed into a wall to cool the room in which it is located. These convectors are typically designed for individual control, although it is possible to install short ducts to enable the unit to cool more than one room. The units can be designed to heat the house as well.

Hydronic cooling systems are usually more expensive and less effective than ducted air systems. As an alternative, an independent chilled water cooling system may be installed with a ducted air system (see Figure 15.4).

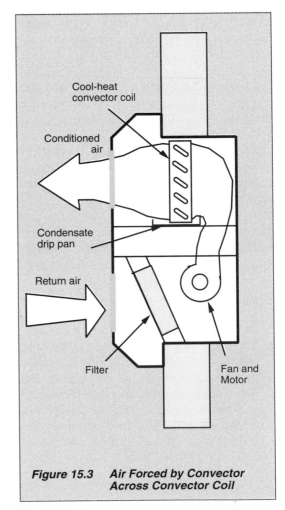

Cool-heat
convector coil

Conditioned
air

Condensate
drip pan

Return air

Filter

Fan and
Motor

Figure 15.3   *Air Forced by Convector Across Convector Coil*

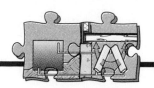

15.3 OTHER APPROACHES

15.3.1 Radiant Cooling Systems

One innovative approach relies on the relatively cool temperature of soil surrounding, or beneath, the house. Water is circulated through a piped subsurface network where it is cooled to ground temperature. Then it is piped to radiant cooling units. These consist of a metal heat collector into which is embedded the cool water circulation system. The cool water absorbs the heat from the metal plate, thus cooling the interior.

15.3.2 Whole-house Fans

In some areas, whole-house fans are used to reduce heat build-up in houses. A fan is installed in the ceiling of the highest level of the house (usually the attic) and connected to a thermostat. When the air temperature reaches a preset level, the fan blows air out of the house, creating a negative pressure. Cooler air then rises up to the upper levels from the basement. The size of the fan required depends on the volume of air that must be exhausted to maintain comfort conditions.

Provision should be made to close off these fans during the heating season to prevent air leakage. If possible, the fans should also be insulated to reduce conductive heat loss into the attic space.

Since whole-house fans significantly depressurize a house, they should not be used in houses with gas- or oil-fired hot water systems that are not directly vented. They are also not recommended where radon gas may be a concern.

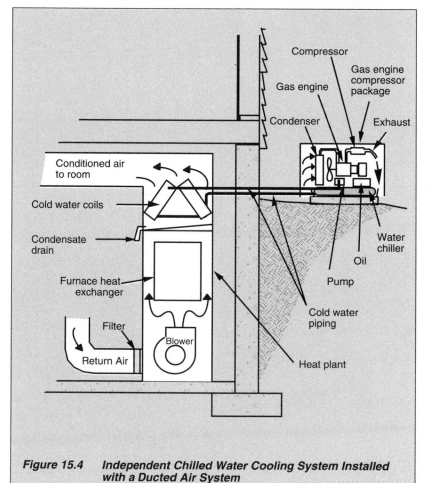

Figure 15.4   Independent Chilled Water Cooling System Installed with a Ducted Air System

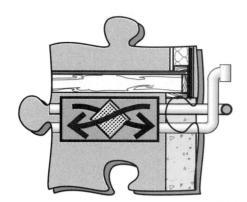

Chapter **16**

VENTILATION SYSTEMS

Ventilation standards throughout Canada are governed by local codes and standards. A considerable amount of work has been conducted in refining these regulations through the 1980s and 1990s. At this point much of the regulatory community has espoused CAN/CSA-F326-M91, *Residential Mechanical Ventilation Systems*, as the guide post for effective ventilation systems. While not all codes reference this standard, it is likely that by 1995, F326 will be referenced in the National Building Code of Canada along with some prescriptive systems. The R-2000 Program requires compliance with F326.

Ventilation standards for new residential buildings require continuous ventilation for each room — or, at a minimum, require the capability to deliver continuous ventilation. As well, exhaust capabilities must be provided in the kitchen and bathroom. Larger exhausting appliances may require their own make-up air. This chapter discusses approaches and equipment that can be used to satisfy various ventilation requirements. For more information on distribution systems, see Chapter 13.

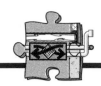

16.1 ELEMENTS OF AN EFFECTIVE VENTILATION SYSTEM

An effective ventilation system has a number of important characteristics.

- It must provide a continuous base level of ventilation sufficient to meet the requirements of the applicable or desired standards.
- Ventilation air should be supplied to all habitable rooms.
- The system should have additional capacity above the base level which can be turned on either manually or automatically when needed.
- In houses with spillage-susceptible, fuel-fired equipment, the ventilation system must not impose imbalances on house pressures which might result in the spillage of combustion gases into the house environment.
- Air exhausted from the building should be taken from those areas where the highest level of water vapour and pollutants is likely (e.g., kitchens, bathrooms, workshops, etc.).
- The fresh air supply should be the cleanest possible.
- The system must be acceptable to the homeowner in terms of comfort and noise levels.
- The ventilation system should be as cost-effective as possible to install and operate.

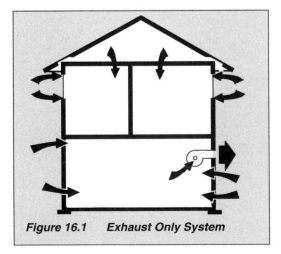

Figure 16.1   Exhaust Only System

16.2 TYPES OF VENTILATION SYSTEMS

There are three basic configurations of mechanical ventilation systems:

- exhaust-only
- supply-only, and
- balanced supply and exhaust.

16.2.1 Exhaust-only Systems

This type of system operates by mechanically exhausting air from the house, using one or more fans (see Figure 16.1). Fresh air enters the house through fresh air inlets and/or through random and uncontrolled cracks and holes in the building envelope. As a negative pressure is exerted in the building from the operation of exhaust appliances, outside air is drawn into the building. Most commonly, outside air is distributed through the ductwork of a forced-air heating system.

While exhaust-only systems are permitted in many jurisdictions, this type of system has several potential disadvantages — particularly in houses with tighter envelopes. Colder air drawn into the house to replace the exhaust air may cause drafts and discomfort unless adequately preheated and mixed with house distribution air prior to delivery to the occupied areas of the home. Some furnace manufacturers have also identified concerns relating to the warranty of their heat exchangers when subjected to colder air entering the furnace return.

Of even greater concern, the negative pressure created by the exhaust fan(s) could cause outside air to be drawn into the building through the chimneys of fireplaces, wood stoves, water heaters and other vented combustion appliances. This could produce spillage or backdrafting — resulting in a potential health hazard.

Exhaust-only systems may also increase the possibility that radon will be drawn from the soil into the house, in areas where radon is found in the soil.

16.2.2 Supply-only Systems

A supply-only system mechanically pushes fresh air into the house, forcing stale air out through chimneys and through random, uncontrolled cracks and openings in the building envelope (see Figure 16.2). The positive pressure created inside the house would minimize the likelihood of back-drafting and ensure that contaminants like radon were kept out of the living space.

However, this positive pressure would force moist household air out of the house through any cracks and holes in the envelope. Decades of actual experience and observation indicate that this approach almost invariably leads to concealed condensation in the walls and attic. As well, the cold fresh air forced into the house during the winter causes discomfort for the occupants unless preheated and properly distributed in small quantities to many locations. For these reasons, a supply-only system is not recommended.

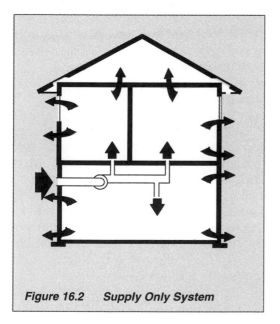

Figure 16.2   Supply Only System

16.2.3 Balanced Supply and Exhaust Systems

The recommended approach to providing ventilation combines balanced exhaust and supply in a single ventilation system (see Figure 16.3). It is then possible to control both the quantity of air exhausted from the space and the quantity of air supplied to the occupants.

A balanced system can be designed to maintain a neutral pressure inside the house, so that there is little pressure difference across the envelope. Each room would have provision for fresh air supply and stale air exhaust either directly in the room or indirectly from an adjoining space. The fresh air is normally distributed to various key living areas of the house through the ductwork of a forced-air heating system or through an independent system of ducts. Stale air is generally exhausted from key odour or contaminant-generating areas such as kitchens, bathrooms and workshops.

A balanced system eliminates most of the disadvantages of an exhaust-only or supply-only system. As with a supply-only system, it is possible, and may be necessary, to heat the incoming air for comfort reasons. However, a balanced system affords the opportunity for exchanging heat from the exhaust air stream to the incoming air stream.

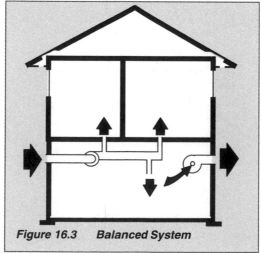

Figure 16.3   Balanced System

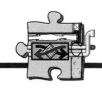

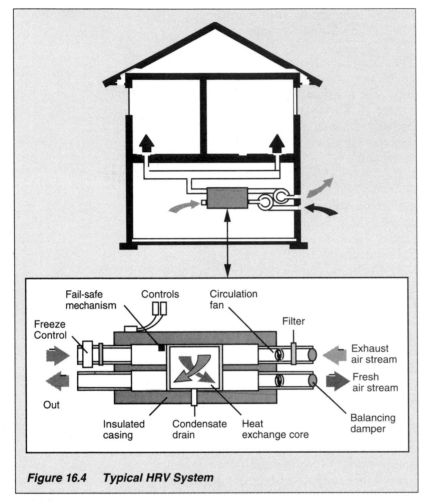

Figure 16.4   Typical HRV System

16.3   HEAT RECOVERY VENTILATORS

Regardless of the type of ventilation system used, warm indoor air is replaced with cold outdoor air which must be heated. One way to reduce the cost of heating the fresh air is to extract heat from the outgoing air and use it to heat the incoming air. Heat recovery ventilators (HRVs) — also referred to as energy recovery ventilators — are devices that do this (see Figure 16.4).

The economics of installing an HRV will depend on the cost of energy and the severity of the weather in a particular location. HRVs can reduce the amount of energy needed to preheat ventilation air, but usually have higher capital costs than ventilation systems without heat recovery.

In addition to the capital costs of the heat exchanger and housing, heat recovery ventilation systems require ductwork to bring the exhaust and supply air to one location for the heat exchange process. Since moist, stale indoor air is being cooled through the HRV, the HRV system must also be able to drain humidity that condenses out of the exhaust air and provide defrosting if necessary.

16.3.1   Types

The heart of an HRV is the heat exchange core, whose function is to transfer heat from the exhaust air to the supply air. Most HRVs are built around "passive" heat exchangers; only heat pump HRVs actively recover heat from the exhaust air. Residential HRVs fall into one of five categories, depending on the type of core used:

- plate (see Figure 16.5)
- rotary or heat wheel (see Figure 16.6)
- capillary blower (see Figure 16.7), and
- heat pump (see Figure 16.8).

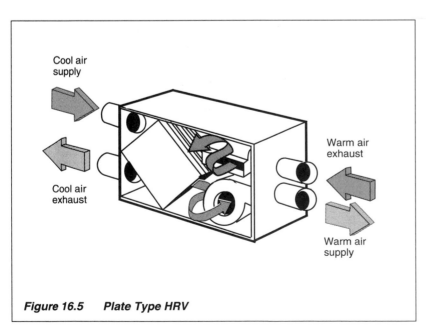

Figure 16.5   Plate Type HRV

The flat plate core is the most commonly used configuration. The core is constructed of thin sheets or plates of metal or plastic that form a series of channels. The supply air and exhaust air pass each other in alternate channels. The plates between the air streams prevent the supply and exhaust air from mixing, while allowing heat to be transferred from the "warm" side to the "cold" side.

The heat exchanger in a rotary HRV consists of a core or wheel that has high thermal inertia—that is, it will only change temperature very slowly. A motor rotates the wheel so that part of it is in the exhaust air stream while the other part is in the supply air stream. As the wheel rotates, it carries heat from the warm (exhaust) air to the cold (supply) air.

A capillary blower HRV combines the function of the HRV fans and the heat exchange core into one fan wheel mechanism. The fan wheel is a porous foam ring which spins rapidly between the exhaust and supply air sections. The spinning wheel flings air from the inside to the outside of the ring while dividers keep the supply and exhaust air streams separated. Heat is transferred from the exhaust air to the supply air section by the foam ring.

Heat pump HRVs use a refrigeration system to extract heat from the outgoing stale air. This heat, plus compressor heat, can be used to heat domestic hot water or to preheat the fresh air. If the heat is not used to preheat the incoming fresh air, you may need to add a separate preheating system. One approach is to use a plenum heater; another is to preheat the fresh air with the return air of the furnace.

In general, the refrigeration portion of heat pump HRV equipment is self-contained, so a refrigeration mechanic is not needed to install the equipment.

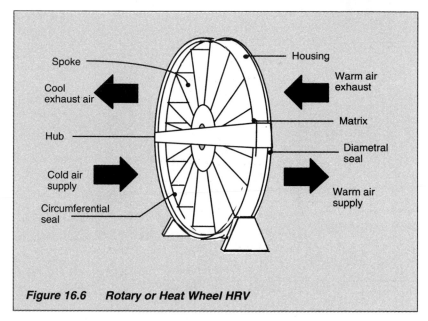

Figure 16.6   Rotary or Heat Wheel HRV

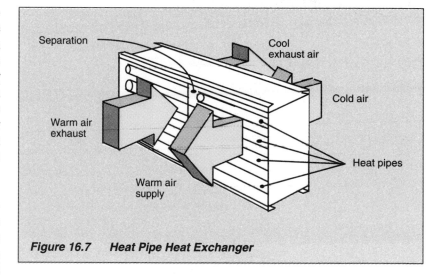

Figure 16.7   Heat Pipe Heat Exchanger

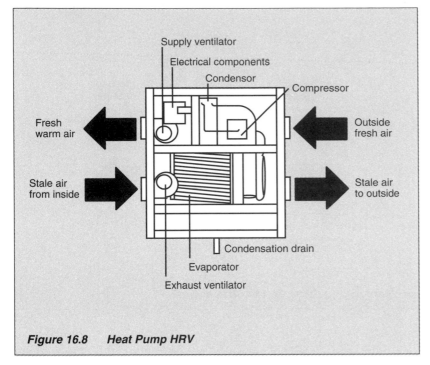

Supply ventilator

Electrical components

Condensor

Compressor

Fresh
warm air

Outside
fresh air

Stale air
from inside

Stale air
to outside

Condensation drain

Evaporator

Exhaust ventilator

Figure 16.8   Heat Pump HRV

16.3.2   Performance Testing and Rating

The performance of an HRV is determined by its air handling capabilities and the percentage of heat it can transfer from the exhaust air to the supply air. HRVs must be rated in accordance with CSA-C439 *Standard Methods of Test for Rating the Performance of Heat Recovery Ventilators*. The standard describes the test equipment, instrumentation, procedures and calculations needed to determine air flows and heat recovery efficiency. These tests are carried out at a variety of air flows and two temperatures: 0°C (32°F) and -25°C (-13°F). Some manufacturers have also begun to test their units at -40°C (-40°F) to determine their effectiveness for northern applications.

The results of these tests are reported on an HRV Design Specification Sheet that you should request from the supplier of the equipment. These sheets are published in the Home Ventilating Institute's *Certified Home Ventilating Products Directory*.

A sample specification sheet is shown in Figure 16.9. The specification sheet is a useful tool, but bear in mind that it should not be the sole basis for selecting an HRV. Cost, warranty provisions, reliability, servicing and suitability of the unit for the climatic zone are some of the other issues that must be considered.

16.3.3   Reading the HRV Design Specification Sheet

You can use the HRV Design Specification Sheets to compare the performance and efficiencies of different HRVs and to determine what units can satisfy the base ventilation requirements of a given house. From the heat recovery efficiency you can estimate the anticipated energy savings from a properly installed and operated unit, relative to a system providing the same air change without heat recovery. Remember that the air flow performance of an HRV is more important than the heat recovery efficiency. A high heat recovery efficiency is of little value if the HRV cannot satisfy the base ventilation requirements.

To determine whether a particular unit will satisfy the ventilation requirements of a given house, compare the efficiency and installation details with other HRVs:

• Determine the maximum airflow needed through the unit at high speed to meet ventilation requirements. Find this number in the "Net Supply Air Flow" column, then look to the left and find the "External Static Pressure" available at this flow. This is a measure of how much duct resistance the unit can overcome at a set flow. If the external static pressure available at the required flow is greater than 100 Pa (0.4 in. of water), or if you know or suspect that the duct resistance will be higher than normal, select a more powerful unit or consult a duct designer.

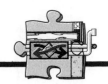

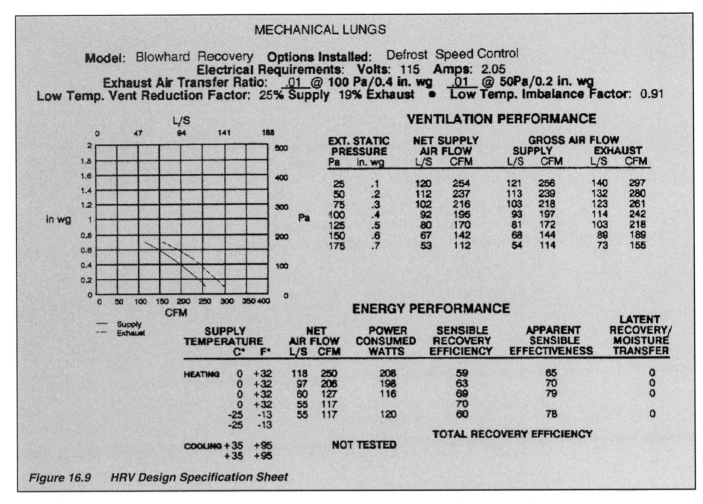

MECHANICAL LUNGS

Model: Blowhard Recovery **Options Installed:** Defrost Speed Control
Electrical Requirements: Volts: 115 **Amps:** 2.05
Exhaust Air Transfer Ratio: .01 @ 100 Pa/0.4 in. wg .01 @ 50Pa/0.2 in. wg
Low Temp. Vent Reduction Factor: 25% Supply 19% Exhaust ● **Low Temp. Imbalance Factor:** 0.91

VENTILATION PERFORMANCE

EXT. STATIC PRESSURE		NET SUPPLY AIR FLOW		GROSS AIR FLOW SUPPLY		EXHAUST	
Pa	in. wg	L/S	CFM	L/S	CFM	L/S	CFM
25	.1	120	254	121	256	140	297
50	.2	112	237	113	239	132	280
75	.3	102	216	103	218	123	261
100	.4	92	196	93	197	114	242
125	.5	80	170	81	172	103	218
150	.6	67	142	68	144	89	189
175	.7	53	112	54	114	73	155

— Supply
-- Exhaust

ENERGY PERFORMANCE

	SUPPLY TEMPERATURE		NET AIR FLOW		POWER CONSUMED	SENSIBLE RECOVERY	APPARENT SENSIBLE	LATENT RECOVERY/ MOISTURE
	C°	F°	L/S	CFM	WATTS	EFFICIENCY	EFFECTIVENESS	TRANSFER
HEATING	0	+32	118	250	208	59	65	0
	0	+32	97	206	198	63	70	0
	0	+32	60	127	116	69	79	0
	0	+32	55	117		70		
	-25	-13	55	117	120	60	78	0
	-25	-13						
					TOTAL RECOVERY EFFICIENCY			
COOLING	+35	+95			**NOT TESTED**			
	+35	+95						

Figure 16.9   HRV Design Specification Sheet

- When comparing the energy performance of different units, look at their "Sensible Recovery Efficiency". This figure provides the basis for a comparison which integrates fan power, defrost and other factors. Efficiencies are given at various flows and at two test temperatures—0°C (32°F) and -25°C (-13°F). Find the air flows that match the continuous flow rate required by the house, then determine the Sensible Recovery Efficiency. The higher the number, the better.

16.3.4   Selecting an HRV

Many different products are available on the marketplace. The selection of an HRV should be based on the following:
- The ability of the equipment to meet the ventilation requirements for the dwelling.
- The heat recovery efficiency of the HRV during cold weather operation.
- The defrost mechanism of the equipment — specifically whether it can depressurize the building and affect make-up air requirements.
- The availability of the equipment and trained service people and installers in your area.
- The installed cost of the equipment.

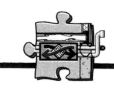

16.4   FANS

Fans are the component of the ventilation system that move the air. An HRV will have the necessary fans built in. If you do not want to use an HRV, you must select appropriate fans for the ventilation system.

Fans consist of a motor and a blade or wheel. They are classified according to their method of operation (see Figure 16.10).

- *Axial fans* move air perpendicular to the impeller or blade. They are efficient at low static pressure, and are normally used in non-ducted applications.

- *Centrifugal fans*, or blowers, have impeller wheels that rotate inside a housing. The air is spun out as the impeller turns. They move air more efficiently at higher static pressures and are most commonly used in ducted applications.

- *In-line fans* are installed within the ducting or within manufactured ventilation equipment. Their ease of installation represents a major benefit. Several manufacturers produce in-line fans capable of dealing with colder air temperatures.

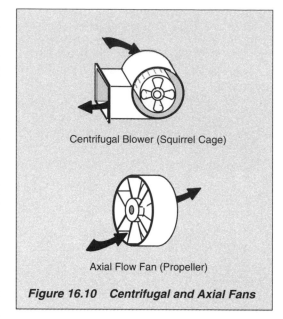

Centrifugal Blower (Squirrel Cage)

Axial Flow Fan (Propeller)

Figure 16.10   Centrifugal and Axial Fans

Fans are rated according to air flow capacity and sound level. Air flow capacity is measured in litres per second (l/s) or cubic feet per minute (cfm) at a certain static pressure. Air flow should be determined based on the rating of the fan at 50 Pa. Sound level is measured in sones or in decibels, depending on the manufacturer. A quiet refrigerator operating in a quiet room produces about one sone. A fan with a sone level of less than two sones will provide acceptable performance.

Sones follow a linear scale: three sones equals three times the noise level of one sone. Decibels follow a logarithmic scale: an increase of 10 dB corresponds approximately to a doubling of the perceived sound level. To ensure quiet operation, fans must be installed in accordance to manufacturer's specifications.

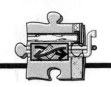

The fan performance curve (see Figure 16.11) provides the relationship between the static pressure across the fan and the air flow it will deliver. Air flow will increase as the external static pressure (the pressure difference between intake and discharge) decreases. Every fan has a characteristic performance curve for each rotational speed.

Figure 16.11 also shows a typical external static pressure curve for a duct system. The operating point is defined as the point where the fan performance curve intersects the external static pressure curve. The fan should be selected to match the operating point. A larger fan will provide more air flow, but will also use more energy.

Fan capacity and sound level ratings should be determined using CSA-C260 *Rating the Performance of Residential Ventilating Equipment*. Most fans tested to this standard will have a Home Ventilating Institute (HVI) certified rating and label.

Fans which handle untempered outside air may be prone to the formation of condensation and frost. Fans designed to handle outside air must be approved by manufacturers for that intended application.

Fans intended for use as the primary ventilator must be capable of continuous operation.

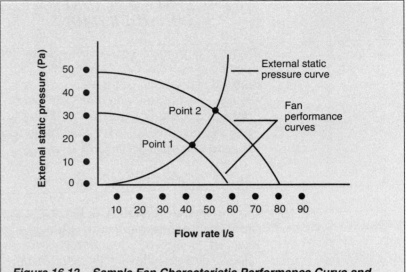

Figure 16.12   Sample Fan Characteristic Performance Curve and External Static Pressure Curve

16.5 SYSTEM PLANNING CONSIDERATIONS

Whatever type of ventilation system you decide to use, careful planning and system design is necessary for a successful installation. Before subcontracting the detailed design and installation of the ventilating system to a certified installer, you must make several important design decisions.

16.5.1 Selecting the Equipment

Before selecting an HRV or fan unit, determine how much ventilation (outdoor) air is needed to meet the requirements for the house design. Ensure that the installed system can meet the supply and exhaust air flow requirements. (For HRVs, refer back to Section 16.3).

Consider also cost, warranty provisions, reliability, servicing and suitability of the unit for the climatic zone.

16.5.2 Locating the Exterior Supply Inlet and Exhaust Outlet

Supply air inlets and exhaust air outlets must be carefully located and installed to avoid contamination and other problems.

Supply Inlet

All fresh air inlets must be located well away from areas of possible contamination. Problem areas include garages and driveways, fan exhaust outlets, vents from combustion appliances, gas meters, oil fill pipes and garbage storage areas.

The air inlet should be located where it will not be blocked. Current installation codes (F326) specify that the air inlet must be at least 450 mm (18 in.) above the finished grade. The inlet must be screened, and must be accessible so that the homeowner can clear out leaves and lint. The screen should be accessible and removable for regular maintenance.

Remember that both the incoming air, and the outgoing air from an HRV, is cold. The cold air ducts must be insulated and covered with a vapour diffusion retarder (VDR) to prevent condensation from forming on the ducts. The VDR on the "cold side" ducts must be effectively sealed to the house VDR (see Figure 16.12). Ducts must also be carefully sealed where they penetrate the air barrier.

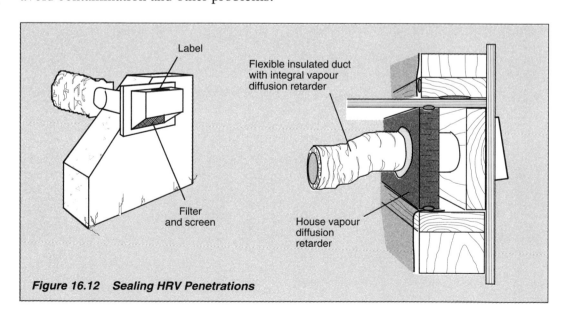

Figure 16.12   Sealing HRV Penetrations

Label

Filter and screen

Flexible insulated duct with integral vapour diffusion retarder

House vapour diffusion retarder

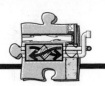

Exhaust Outlet

The exhaust outlet should not be located where it could contaminate fresh incoming air, nor in attics, garages or on walkways where condensation, moisture or ice could create problems. According to the CSA installation standard, it should be at least 1.8 m (72 in.) to the side of, above, or below the supply inlet (see Figure 16.13). The bottom of the hood must be a minimum of 100 mm (4 in.) above grade or other horizontal surfaces. The exhaust duct must be carefully sealed where it penetrates the air barrier.

16.5.3 Locating the Ventilator

In determining the location of the ventilating unit itself, you need to consider such things as the inlet and outlet locations, noise levels, drainage, power supply, etc.

- The unit should be located in the heated interior, away from noise-sensitive living areas (dining room, living room, bedroom, etc.). Vibration isolation of the fans should be provided to minimize impact noises in the dwelling.
- It should also be: close to an outside wall to minimize insulated duct runs; centrally located to optimize the distribution system; near a drain for condensate disposal, and close to an electrical supply.
- The unit should also be easily accessible for maintenance purposes, and should be located well away from hot chimneys, electrical panels and other possible fire sources.
- Try to minimize the length of distribution ducting required, particularly for HRVs. Adding duct length adds resistance and reduces air flow.

16.5.4 Air Distribution

Each room must have provision for both fresh air supply and stale air exhaust. Fresh air must be distributed to all "living areas" within the house.

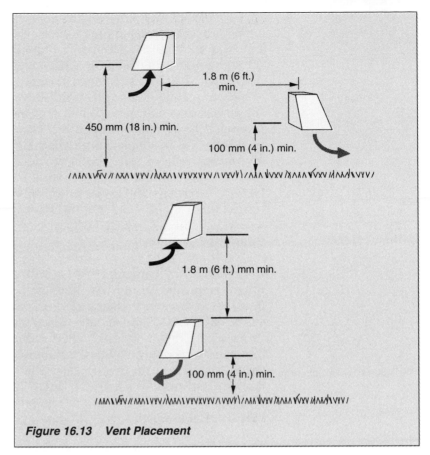

Figure 16.13   Vent Placement

Supply Air

Supply air can either be distributed through an independent duct system or integrated with, and distributed through, the ductwork of a forced-air heating system.

Independent Systems

This approach must be used in houses that do not have a forced-air distribution system. Using independent ducts, the temperature of the air supplied to each room may be below room temperature. For example, if it is -10°C (14°F) outside, an HRV may only warm the supply air up to 10°C (50°F) unless there is an auxiliary duct heater.

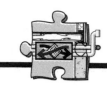

The supply air will behave like cool air supplied by an air conditioning system. The best way to supply cool air to a room is through a high wall or ceiling outlet which discharges air horizontally along the ceiling —like an air conditioning diffuser. Incoming air has time and space to mix with the room air before dropping down to the "occupied zone". If the supply air must be delivered through a floor grille, the air should be directed up the wall (see Figure 16.14 — bottom). Undersizing of grilles may result in restrictions to air flow and resulting noise.

Combined Systems

Combining the ventilation system with a forced-air heating system can eliminate the need for separate supply ductwork. Fresh air can be delivered to the cold air return of the furnace (see Figure 16.15). The furnace blower must operate continuously, on a lower speed, to distribute the fresh air throughout the home and mix it with house air. In colder climates, high wall diffusers may still provide acceptable comfort.

Exhaust Air

Exhaust air grilles should be located in the rooms where the most water vapour, odours and contaminants are produced (e.g., kitchens, bathrooms, laundry rooms and workshop areas). Rooms with only exhaust grilles will receive fresh air only indirectly, from other areas of the house. Undercut doors or install transfer grilles to provide adequate air flow to these rooms (see Figure 16.16).

Exhaust grilles in kitchens must be located in the ceiling or on the wall within 300 mm (12 in.) of the ceiling.

16.5.5   Dampers

Dampers can be adjusted to determine how much air will flow through the system. Of greater importance in exhaust-only systems, mechanical dampers can regulate the entry of outside air into the building — as determined by operation of the exhaust appliances.

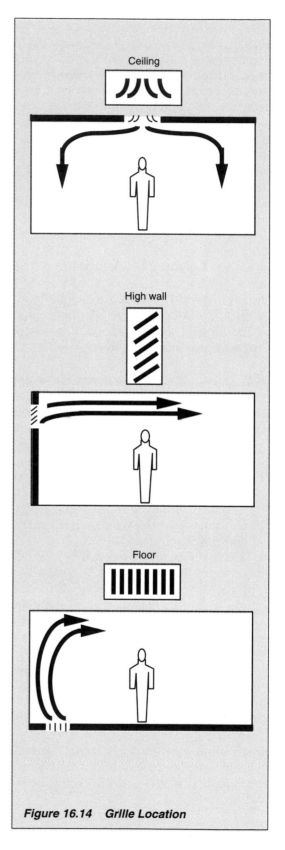

Figure 16.14   Grille Location

Similarly, mechanical dampers can be wired to open upon operation of any single large capacity exhaust appliance — such as a range barbecue. This can help control make-up air to reduce potential pressure imbalances.

Backdraft dampers should be provided on all exhaust hoods to prevent air leakage when the fans are not in operation. Poorly designed dampers can freeze open or shut in cold weather. Dampers must be capable of free operation at all times.

16.5.6   Controls

Most residential ventilation systems are designed to operate at two speeds. At low speed, they provide the required base level of ventilation continuously. At high speed, they can deal with extra contamination.

A high speed control can be activated either manually or by a a humidistat. Once the indoor humidity rises above a preset level, the ventilation system switches to high speed. A manual crank timer type switch is preferable to a pure on/off switch because there is less chance the system will be left on high speed when it is not really necessary.

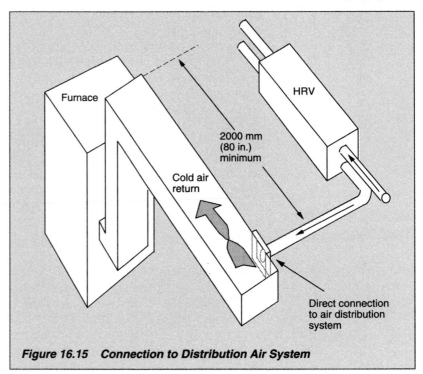

Figure 16.15   Connection to Distribution Air System

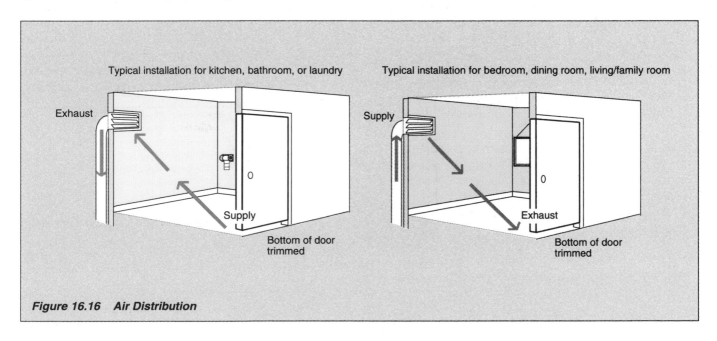

Figure 16.16   Air Distribution

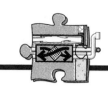

Kitchen and bathroom fans are normally controlled with manual switches, but can be connected to timers. This is particularly useful for bathroom fans, since water vapour generated by showers must usually be exhausted for several minutes after the shower is no longer in use.

16.5.7   Air Flow Measuring Stations

Air flow measuring stations should be installed in the distribution ducts to enable the installer to balance the system (see Figure 6.17). The stations should be located so that all the supply and exhaust air is measured. All joints between the measuring stations and the exterior should be sealed. If the ventilation system includes an HRV, the air flow measuring stations should be installed on its warm side (see Figure 16.18).

Before any air flow measurements are conducted, air should be blown through the measuring station to clean out dust and lint. Air flow readings should be taken when the system is operating in the normal continuous ventilation mode. Final measurements should not be taken until final system adjustments have been made.

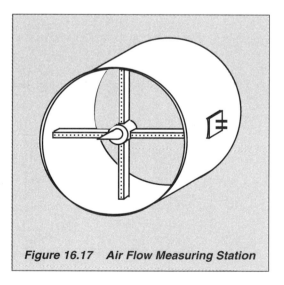

Figure 16.17   Air Flow Measuring Station

16.6   SYSTEM INSTALLATION

Where possible, hire a mechanical contractor who has passed the Heating, Refrigerating and Air Conditioning Institute of Canada (HRAI) *Residential Mechanical Ventilation Design and Installation Course*. Ensure that the subcontractor complies with all design and installation requirements, balances the system and fills out any required installation reports. In supervising and inspecting the installation, use the checklist provided in Appendix III.

The mechanical subcontractor should measure air flows, and balance the ventilation and exhaust air systems. Balancing air distribution systems will involve adjusting fans and dampers to ensure that the desired quantities of air are delivered to, and exhausted from, each area of the house. The ventilation supply fans must bring the same amount of fresh air into the house as the exhaust fans are removing. To make it easier for the installer to balance the system, dampers should be located in an accessible location near the main supply/exhaust grilles or in the unfinished basement or utility room.

For HRVs, the installing contractor should complete CSA Standard F326 Form A, which describes the installation, provides air flow measurement results, and ensures that the HRV system meets industry standards. The mechanical subcontractor should provide a written air balance report on the ventilation and exhaust air system.

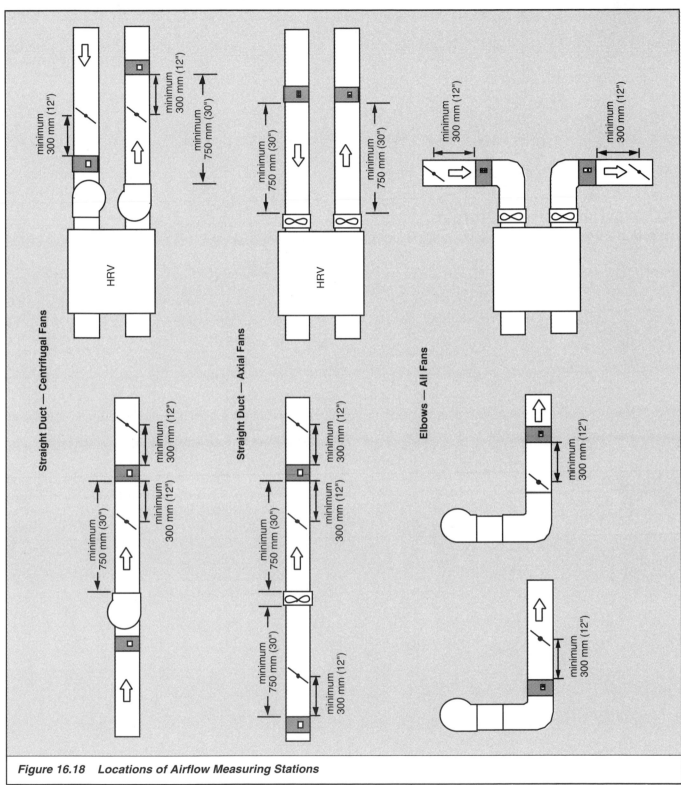

Figure 16.18   Locations of Airflow Measuring Stations

Chapter **17**

DOMESTIC HOT (TAP) WATER (DHW) SYSTEMS

In energy efficient houses, some water heaters may consume more energy than space heating or lights and appliances. This chapter looks at how you can improve the efficiency of the domestic hot water (DHW) system by selecting and properly installing efficient equipment. Section 17.1 reviews some system design and operation considerations. Sections 17.2 through 17.7 look at the different types of water heating equipment that are available.

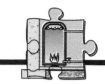

17.1 REDUCING ENERGY LOSSES

Domestic hot water can be heated in several ways.

- Through a central water heater.

- Through instantaneous or tankless heaters.

- Through integration with the same heating system.

The heat can be supplied from an oil or gas burner, an electric resistance element, from an earth energy heat pump, or even from solar panels.

The operating efficiency of a domestic hot water system can be significantly improved by designing the system and selecting equipment to reduce standby and line losses, and by installing devices to limit hot water wastage.

17.1.1 Reducing Standby Losses

The term "standby loss" refers to heat lost to the surrounding air from the water in a central domestic water heater and its distribution system. It is proportional to the temperature difference between the water and the surrounding air, the surface area of the tank and pipes, and the amount of insulation encasing the tank and pipes. There are several steps you can take to reduce standby losses.

- Place the DHW heater over a layer of rigid insulation to reduce heat loss through the bottom of the tank.

- Install a heat trap above the water heater (see Section 17.1.2)

- Increase the level of tank insulation.

- Insulate the hot water pipes (see Section 17.1.3).

17.1.2 Heat Traps

As noted in Chapter 2, convective loops will form in a fluid when temperature differences are present. The warmer fluid will rise to the top and the cooler material will fall. This loop will continue until all of the fluid is at the same temperature.

This type of convective loop will form in a water heater, especially if it sits on a cold concrete basement floor. A convective loop will also form in the pipe that supplies hot water to the house, where heat will dissipate to the interior air. The water in the pipe will cool quite quickly because the surface area of the pipe is large relative to the volume of water. As a result, considerable energy can be wasted because of radiation from this pipe.

A heat trap is a simple piping arrangement that prevents hot water from rising up the pipe in the first place, thereby reducing heat loss (see Figure 17.1).

17.1.3 Reducing Pipe Losses

The best way to reduce heat loss from the piping is to minimize the length of hot water piping required. Even when this is done it is a good idea to insulate all hot water pipes to further reduce heat loss. Pipe insulation is available in a variety of materials and thicknesses, with easy application to most water pipes. Use insulation with a minimum RSI of 0.35 (R-2) over at least the first metre of pipe from the tank.

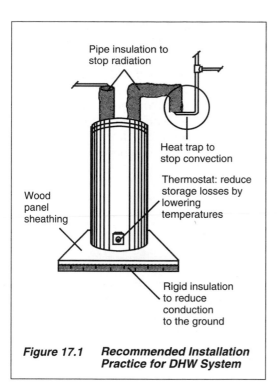

Figure 17.1 **Recommended Installation Practice for DHW System**

- Pipe insulation to stop radiation
- Heat trap to stop convection
- Thermostat: reduce storage losses by lowering temperatures
- Wood panel sheathing
- Rigid insulation to reduce conduction to the ground

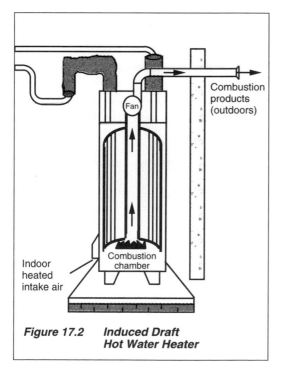

Figure 17.2 **Induced Draft Hot Water Heater**

- Fan
- Combustion products (outdoors)
- Indoor heated intake air
- Combustion chamber

17.2 GAS-FIRED WATER HEATERS

There is a wide range of gas-fired water heating equipment available today, ranging from conventional, naturally aspirating, natural draft, gas-fired water heaters with a seasonal efficiency in the range of 55%, to high efficiency condensing units which can have an efficiency of 90%. A measure of seasonal efficiency of water heaters is given by their "energy factor". It considers standby losses (affected by the amount of tank insulation), and off-cycle vent losses (affected by the burner and heat exchanger). The higher the energy factor, the more efficient the water heater. Some provinces are setting minimum efficiency levels for water heaters.

17.2.1 Natural Draft Gas-Fired Water Heaters

These units are quite cheap to make and do not have any fan, either upstream or downstream of the unit, to help move the combustion products. There is a naturally aspirating burner and a draft hood, as with a conventional gas furnace. They do require a vertical chimney (vent). The chimney/vent creates a natural draft only, and draft losses can be quite significant. As a result, they are more subject to combustion spillage than other gas-fired equipment. Their overall efficiency is usually fairly low, with an energy factor in the low 50's. In a tight house, it might be advisable to consider one of the alternative gas-fired water heaters described below, either direct-vented or with a powered exhaust.

17.2.1 Induced Draft (Fan-Assisted) Gas-Fired Water Heaters

Induced draft hot water heaters (see Figure 17.2) use fans downstream of the water heater proper to move combustion by-products through the venting system rather than relying on dilution air to move gases by natural convection. They can be side-wall vented and thus are more compatible with high efficiency condensing furnaces. Many of these units still have a pilot light and a draft-hood, tending to lower their overall efficiency.

If side-wall venting the water heater, it is important to install the unit with a vent connector always sloping upward from the water heater to the outside. Care should be taken to use the minimum equivalent length (EL) of vent connector possible, to minimize the chance of condensation in the venting system.

Most induced draft (fan assisted) units require combustion air and dilution air from inside the house. In a poorly sealed building, outside air will leak into the house through cracks and holes in the building envelope to replace the air used for combustion. In a well built house, a source of combustion air must be provided near the water heater. If you use a unit that does not come with a built-in, directly connected combustion air intake, you must provide an inlet sized and located according to the Canadian Gas Association (CGA) standard CAN/CGA-B149.1-M86.

If there is another gas-fired appliance such as a furnace located in close proximity to the DHW heater, only one combustion air supply is needed. The inlet should be sized to meet the requirements of both appliances, based on the total fuel burning capacity of the two burners.

17.2.2 Direct Vent Gas-Fired Water Heaters

Direct vent, or sealed combustion, water heaters are designed to bring all of their combustion air in from outdoors through a directly connected inlet, and to vent combustion by-products directly outdoors. There is no interaction between the appliance and the house air (see Figure 17.3).

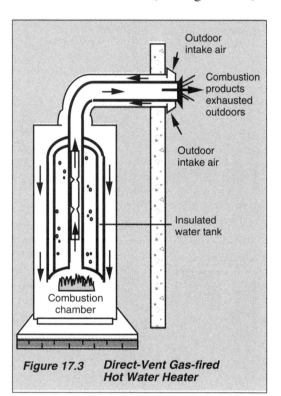

Figure 17.3   Direct-Vent Gas-fired Hot Water Heater

17.2.4 Instantaneous Gas-Fired Water Heaters

There are a number of instantaneous water heaters on the market, which have no storage tank, but merely heat up the water as it is required. Thus there is no true "standby loss". Some of the units are direct-vented types. Others operate as side-wall venters, but still require house air for combustion and dilution. In order to supply the quantity of hot water required by North Americans, the firing rates must be very high, so that air requirements (if applicable) and thermal stressing of the heat exchanger are important factors.

17.2.2 Condensing Gas-Fired Water Heaters

Gas-fired condensing water heaters, similar to the high efficiency condensing furnaces described in Section 14.2.2, recover the latent heat in the flue by using the cold mains as the forcing function for condensation. They only need a small plastic pipe to exhaust the combustion products out through the side wall of the house. Some of these units can achieve efficiencies as high as 90%.

In many new designs, *integrating the space and water heating* applications in one unit can allow condensing high efficiencies for both applications, as discussed in Section 18.2.1.

17.3 OIL-FIRED WATER HEATERS

17.3.1 Stand-Alone Oil-Fired Water Heaters

Stand-alone, oil-fired water heaters use some form of flame retention head burner (see Section 14.3.1) as the heat source, directly heating the water in a storage tank. Efficiency gains have been made by baffling if the flue passage to increase heat transfer and by better insulation of the tank. Equipment having a high static burner realizes additional gains, due to both more efficient energy generation and reduced off-cycle losses up the chimney, because of the high pressure drop across the burner head.

Most of the oil-fired water heaters can use the same venting system as the oil-fired central furnace/boiler, providing the flue size is adequate to carry all the combustion products.

Some of the oil-fired water heaters have an additional induced draft fan mounted on the wall, which allows side-wall venting of the appliance.

17.3.2 Tankless Coil Water Heaters

In the Maritimes, tap water heating is often done by a tankless coil installed inside an oil-fired hot water boiler. This system can be quite inefficient, requiring the boiler to be hot at all times, even during the summer. Inefficient short-cycling is common, as is the potential for high boiler casing losses and off-cycle stack losses.

17.3.3 Oil-Fired Boiler with External Integrated Tap Water Storage

Using a low-mass boiler fired with a high static retention head burner, coupled to a well-insulated tap water storage tank by an efficient water-to-water heat exchanger results in efficient water heating, as well as home heating. When the thermostat calls for heat, the boiler runs normally. When the thermostat is satisfied, the burner continues to run instead of shutting off, but the hot water from the boiler is passed through the heat exchanger, sending heat to the storage tank instead of the house. If hot tap water is required, it is taken directly out of the storage tank. More details of this system are described in Section 18.2.1.

17.4 ELECTRIC WATER HEATERS

17.4.1 Electric Resistance Water Heaters

Electric resistance water heaters (usually 40 gallons capacity) are widely used, readily available, inexpensive to buy, and simple to install. They should have a factory installed insulation level of at least RSI 1.75 (R-10). The seasonal efficiency of these units is 93 percent.

In houses where large amounts of hot water are used, electric DHW tanks may be less effective than fuel-fired units since they have a slower recovery time. This problem can be overcome by installing a larger volume water tank.

However, although the initial cost of an electric water heater is very low, its cost of operation can be significantly higher than that of oil- or gas-fired water heaters, due to the higher cost of electricity in most areas of Canada. The exceptions are those areas where electrical off-peak reduced rates are available. In most areas of Canada, efficient fuel-fired water heaters may be the best option, as they have a much faster recovery time, as well as cheaper operating costs.

17.4.2   Heat Pump Water Heaters

Many of the heat pumps used for space heating can also be used to heat water. In addition, there are heat pumps on the market designed to provide only hot water (see Figure 17.4); they extract heat from the inside air or the exhaust airstream of the house to heat the water. These heat pump water heaters follow the same principles as space conditioning heat pumps (see Chapter 14) and are available either as self-contained units or as packages which can be added on to conventional water heaters.

A heat pump will supply more energy than it uses: the ratio of energy supplied to energy used is defined as the coefficient of performance (COP). Heat pump water heaters generally operate with a COP in the range of 2.5 to 3, and can save up to 60 percent of the energy used by a typical water heating system.

During the summer, a heat pump water heater can reduce air conditioning requirements if it extracts heat from the inside air. However, if it extracts heat instead from the exhaust airstream of the ventilation system, no summer air conditioning effect will be obtained without additional equipment. Conversely, in the winter if the heat is obtained from the heated indoor space, there will be an increased load on the space heating system.

Because of the high price and long payback period, heat pump water heaters have not been very popular.

17.4.3   Earth Energy Systems as Water Heaters

Most Earth Energy heat pump systems have the capability of heating domestic hot water. While in air conditioning mode in the summer, an Earth Energy system will remove heat from the house and transfer it to the hot water storage tank. It can provide practically free hot water in the summer, and hot water for approximately one-third of the price of a normal electric water heater during the heating season. Such operation is analogous to the operation of other integrated space and water heating systems, as discussed in Chapter 18.

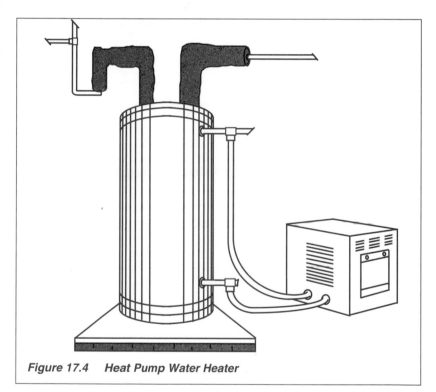

Figure 17.4   Heat Pump Water Heater

17.5   SOLAR WATER HEATERS

Solar DHW systems have been on the market for several years (see Figure 17.5). Most commercial systems can provide about 40 to 50 percent of a family's annual domestic hot water needs, although in some cases they can provide more. (Note: these are pre-heat systems only; a conventional tank is still required.) These systems may be economically advantageous in some locations — specifically in areas of high energy costs and considerable available sunshine.

The installation of a solar water heater must be carefully considered at the design stage. An area of unshaded southern exposure is needed for installation of the solar collectors.

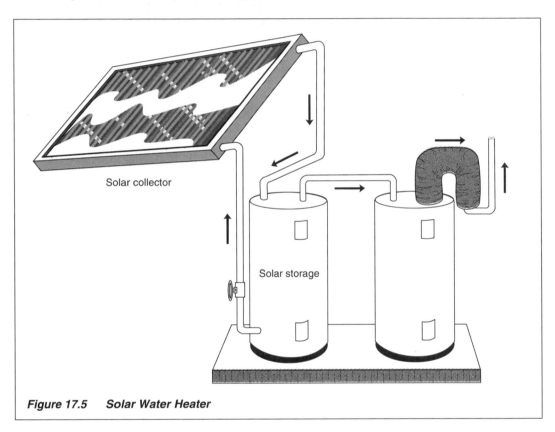

Figure 17.5   Solar Water Heater

Chapter **18**

INTEGRATED MECHANICAL SYSTEMS

This chapter explores integrated mechanical systems (IMS) including an examination of their benefits, a description of systems which have been successfully employed in the past and are readily available now, and a look at where developmental work is likely to take such systems in the future.

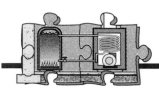

18.1   BACKGROUND

The term "integrated mechanical system" generally refers to a system which replaces most, if not all, of the functions normally provided by separate pieces of mechanical equipment such as the furnace, hot water heater, heat recovery ventilator and air conditioner. However, the few complete systems in existence are still in the experimental stage. For the purposes of this chapter, an IMS is defined as any system which synthesizes the operation of two or more pieces of mechanical equipment. These technologies are plentiful, and are laying the groundwork for systems which will, in the near future, link all the thermal energy flows in a house.

There are a number of reasons why research and development on integrated systems is expanding so rapidly right now.

- Improvements to the building envelope have reduced the space heating load to the point where, in highly energy efficient homes, it is sometimes difficult to justify the expense of a high efficiency furnace solely to satisfy the heating load. In order to take advantage of the efficiency potential of condensing gas-fired systems, it makes sense to combine space heating with other functions.

- Domestic hot water (DHW) loads have remained fairly constant or have even increased over time, making it logical to improve the efficiency of the hot water generator. The natural fit with space heating generation promotes combined space/water heating systems.

- Oil-fired heating equipment cannot operate efficiently at the low firing rates needed in energy efficient homes. Combining the space and water heating applications resolves this issue.

From the builder's perspective, there are a number of other reasons for IMS which also make sound business and design sense. Combining the operation of mechanical equipment can:

- decrease the number of holes to be made in the exterior building structure for fuel and air supply and exhaust;

- decrease the capital cost through reduction in ductwork, number of motors, blowers, burners, casings, cabinets, etc.;

- decrease installation costs as fewer trades are required for less time to install fewer individual pieces;

- decrease the number of electronic controls which have to be set up to run individual pieces of equipment while not infringing on the operation of other equipment in the house;

- benefit the home buyer through reduced electrical and fuel energy operating costs, increased available space and improved, environmentally-responsible home operation: the lower energy used represents significantly less pollution generated.

18.2   HARDWARE APPROACHES

Hardware approaches refer to integrated mechanical systems which combine the function of just two pieces of mechanical equipment. In the past, R-2000 builders constructed these systems themselves using conventional HVAC equipment and the services of qualified HVAC designers and installers with experience in R-2000 construction. Many manufacturers of HVAC equipment are now marketing integrated systems like those explored in this section.

18.2.1   Integrating Space Heating and Tap Hot Water

The integration of space and tap water heating is a logical combination, bringing together the two major energy consumers in the house in a way which reduces the capital cost while synergistically increasing the energy efficiency of both applications, often significantly.

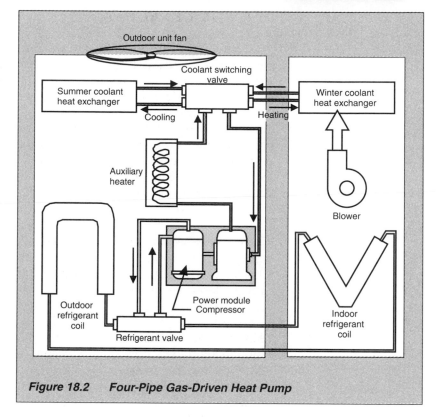

Figure 18.2   Four-Pipe Gas-Driven Heat Pump

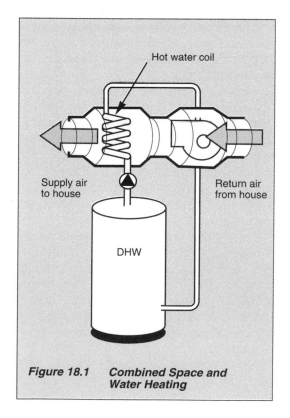

Figure 18.1   Combined Space and Water Heating

Gas-Fired Combined Space/Water Systems

As described in Section 17.2.1, conventional fossil fuel-fired water heaters are fairly inefficient, typically at the 55% level. For a number of reasons, the majority of homes heated with natural gas also have gas water heating.

Combining the space and water heating applications into one integrated, high efficiency, condensing appliance is an ideal application of the IMS principle. The cold mains water temperature and the cold air return temperature provide good driving forces to condense the flue gases and achieve high levels of efficiency for both applications, potentially well over 90%.

When installing such systems, care must be taken to allow the high efficiency potential to be reached. Some of the demonstrations of this technology have gone countercurrent to optimal system design.

For example, one system attempted to passively preheat the mains water before it reaches the appliance. This is not desirable, as the higher water temperature results in less flue gas condensation and lower system efficiency. Sophisticated design and control strategies are now being developed by CANMET to determine the optimal firing rate, storage capacity and stored water stratification temperatures to maximize efficiency, while ensuring that all demands can be satisfied.

Condensing gas-fired boilers in hydronic heating systems can have difficulty condensing in practice, because the return water temperature is above the dewpoint of the flue gases. By installing a water-to-water heat exchanger and storage tank upstream of the boiler, as shown previously in Figure 14.4 the return water temperature can be brought below the dewpoint, flue gases will condense and the efficiencies will be improved significantly.

Mid-efficiency gas-fired combined systems also exist, but their overall efficiency potential is lower than for the condensing ones. Some of these have even used a conventional gas-fired water heater as the basic energy generator, with consequent low overall efficiencies.

Oil-Fired Combined Space/Water Systems

Combining space and water heating in oil-fired systems can also result in high performance and customer satisfaction.

As mentioned in Section 14.3.4, a significant problem exists with oil heating systems in new housing. Because of the limitations in burner technology, firing rates cannot be reduced below 0.5 US gph (70 kBtu/hr); well-insulated new homes can have a design heat load of one-third to one-half this level. Using the IMS systems can allow efficient operation of the furnace/boiler without short cycling and the attendant losses in efficiency and comfort.

One effective oil system couples a mid-efficiency, low thermal mass boiler, fired with a high-static burner, to a well-insulated DHW water storage tank, using an efficient water-to-water heat exchanger. When the house thermostat calls for heat, the boiler turns on, supplying space heating either directly into a hydronic system or through a fan-coil into a warm air distribution system. When the house thermostat demand is satisfied, the boiler, instead off shutting off, continues to run, but dumps the heat across the heat exchanger into the tap water storage tank, as shown previously in Figure 14.4. An oil-fired water heater can also serve as the prime energy source, giving a first-cost advantage, but often at the price of lower system efficiencies. For either system, burner "on" time per cycle is increased, so the boiler tends to run closer to its efficient equilibrium condition.

A variation on this concept uses intelligent microprocessor controls and a reversible pumping system to make the heat flow go in both directions. The next time the house needs heat, the boiler does not need to come on again. Instead, the necessary heat is taken out of the storage tank across the water-to-water heat exchanger and distributed around the house.

If tap water is required at any time, it is merely drawn directly out of the storage tank. If properly installed and controlled, the efficiency for both space and tap water heating can be more than 85%. Even greater reductions in pollutant emissions would result, due to fewer start-ups and shutdowns.

Sophisticated design and control strategies are now being developed by CANMET for these systems, to determine the optimal firing rate, storage capacity and variable stored water stratification temperatures, in order to maximize efficiency, while ensuring that all demands can be satisfied. In such a fashion, oil can indeed be an effective energy source for new housing.

The tankless coil oil-fired boiler IMS system has been common in the Maritimes for more than 20 years. In a hydronically-heated house, tap water is heated by a coil immersed inside the water jacket of a boiler. There is no storage tank. In order to have hot tap water when needed, the boiler must be hot at all times, even during the summer. To be able to supply enough hot water, the units may need very high heat inputs, with firing rates of as much as 1.5 US gph (210 000 Btu). Short-cycling is common, as is the potential for high boiler casing losses and off-cycle stack losses. The tankless coil system is usually very inefficient (although a low thermal mass boiler and well-insulated storage tank can improve performance). Generally, it should be avoided, in favour of those systems described in the preceding paragraphs.

18.2.2 Integrated Heating and Cooling

Heat Pumps as Integrated Systems

Both air-to-air and earth energy heat pumps can provide both heating and cooling capabilities. When heating, the air and earth energy heat pumps take low grade heat from the outside air or ground (water), respectively, and upgrade it to heat the house, as described in Sections 14.4.3.1 and 14.4.3.2.

The operating cycle of both systems can be reversed during the summer months, allowing the heat pumps to act as cooling systems, where they remove the heat from the house air and move it outside to the heat sink — the air or the ground (water). Ground source heat pumps can also be configured to supply a portion or even all of the domestic hot water requirements.

While heat pumps, particularly earth energy systems, are much more expensive than electric resistance heating, they may offer economic value over the long term, especially when space cooling and domestic hot water heating are also part of the system.

There is a significant amount of R&D being carried out at this time on both types of heat pumps, leading to increased efficiencies and reduced prices and operating costs. Indeed, with some experimental earth energy systems, COP's of as high as 8 have been achieved.

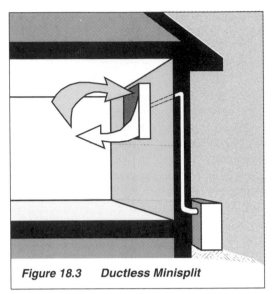

Figure 18.3   Ductless Minisplit

Split Systems

A conventional ducted split system has an indoor section sized to handle the house loads. Distribution ducts carry heated or cooled air to and from individual rooms or zones.

The "ductless mini-split" air conditioner and heat pump is a new type of heating and cooling system which has come from Japan into the Canadian marketplace. This system is divided into an outdoor section, which contains the compressor and the condensing coil, and an indoor section with the evaporator coil and blower. Indoor and outdoor sections are connected by copper refrigerant lines and control wiring.

The "ductless" system is generally sized to heat or cool only a single zone. The indoor section is mounted on a wall or floor in the living space, eliminating the need for distribution ducts.

At their current cost of roughly $2000 per ton of cooling, the mini-splits are probably too expensive to compete with conventional ducted heating and cooling systems for new house construction. However, they might find a market in multi-family construction. Developments to reduce their costs might expand their potential market in the future.

Radiant Floor Heating and Cooling

Radiant floor heating has been discussed in Section 13.3. The concept of radiant floor cooling is quite new and is still undergoing development. There are a number of prototypes. While proponents claim that, like radiant floor heating systems (a claim that itself is debatable), radiant floor cooling provides better comfort and uses less energy than forced air systems, this is yet unproven.

18.2.3   Integrated Heating and Ventilation

With a system combining heating and ventilating capabilities, fresh air can be supplied directly to the forced air distribution section of the home while being tempered either by a heat recovery ventilator (HRV), by a preheater or even by the furnace itself. The space heating ductwork then serves the dual purpose of distributing heat and fresh air.

As discussed in Section 13.1.1, there is some concern that ducted untempered outside air in the return air duct of a mid-efficiency gas furnace could cause premature corrosion of the furnace. This should not be a problem for condensing gas furnaces, or for most oil furnaces.

Certainly, a heat recovery ventilator (HRV) by itself can be considered a form of integrated system, combining the function of ventilation with the supply of some of the heat demand, while making use of the home's heat distribution network.

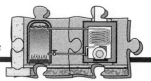

18.3 SYSTEM APPROACHES TO INTEGRATED MECHANICAL SYSTEMS

The more components of the mechanical system that can be combined into one package, the more benefits of integration can be realized. System approaches to IMS can combine most, if not all, of the major mechanical components/requirements.

Attempts in the past to manufacture and market totally integrated mechanical systems have not met with success. Technological problems and the complexities of some systems kept them from gaining widespread acceptance within the building community. However, advances in the technology and control systems are generating a resurgence of innovative R&D in this area.

Some of the systems discussed in this section are commercially available, or can be assembled from readily available components.

18.3.1 Integrated Space Heating, Hot Water and Ventilation

This is the most common "next step" in the integration of major mechanical components. Presently, most of these systems are based around heat recovery ventilator concepts. Such units extract heat from the exhaust air to heat the hot water tank and the incoming fresh air. Additional space heat can be supplied with an electric heating coil, fossil fuel-fired furnace, or even an advanced combustion wood burning appliance.

Given that highly-insulated houses have a low cooling load, which can largely be met using ambient air, very few designers of upcoming systems plan on integrating a cooling capability into their system, with the exception of those who are using a heat pump as the basic energy source.

Combined Space Heating and DHW with HRV Central Supply and Exhaust

Such a system can consist of a high efficiency fuel-fired boiler or water heater which supplies heat to the house through a fan coil in a forced air distribution system, drawing heat from hot water from a well-insulated tank (see Figure 18.4). An HRV is incorporated into the central distribution system and the exhaust ventilation system. As described in Section 18.2.1, space and DHW heating are interactive, synergistically increasing the efficiency of both. The forced air circulation integrates both supply air and heating.

If using this configuration, in addition to following the guidelines of Section 18.2.1, the builder should ensure that the combustion and/or storage units are adequately sized to provide both space and DHW heating under design conditions.

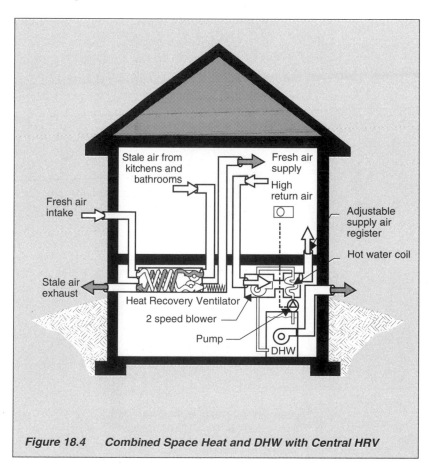

Figure 18.4   Combined Space Heat and DHW with Central HRV

Integrated Heating, Cooling and DHW

As noted in Section 18.2.2, both air-to-air and earth energy heat pumps have the capability of providing integrated heating, cooling including dehumidification and hot water. In the past, domestic hot water heating has been provided by extracting waste heat from the refrigerant during the cooling mode. Models are now available with an improved refrigerant circuit that provides DHW heating in any mode, and even if the system is not running for cooling or heating purposes. Such heat pumps can provide a much higher hot water heating efficiency than can conventional electric water heaters.

Some of the heat pump water heaters (see Section 17.4.2) can combine space heating and cooling with DHW heating. Such units can provide only water heating when there is no space cooling or heating load. When the heat pump water heater cannot satisfy both space and water heating loads, an electric element in the water heater, or an efficient fuel-fired water heater can provide the back-up heat required.

Figure 18.5   Independent Ventilation, Space Heat and DHW

There are some Swedish systems which combine an exhaust-only ventilation system with a heat pump, and use a well-insulated hot water storage tank as a heat sink. These have formed the basis for some experimental systems integrating all the functions of ventilation, heat recovery, water heating and space heating and cooling.

Baseboard Heaters with Central Supply and Exhaust HRV

This is a primitive version of an integrated system, consisting of baseboard heaters controlled by individual wall-mounted thermostats, an electrically-heated DHW tank and a heat recovery ventilator incorporated into a central supply and exhaust ventilation system. Figure 18.5 shows the equipment fans and dampers, controls and direction of air flows. In this case, each component acts independently.

If using such a configuration, the builder should consider the following:

- Supply air in this system is preheated by the exhaust air through the HRV. At certain times of the year, the supply air may not be warm enough. The fresh air supply should be located at the top of the room, or floor registers should point towards the walls. An alternative is to supply air to a dropped ceiling plenum, thereby allowing adequate air mixing.

- An in-duct heater may be required to pre-heat the fresh air supply.

- Exhaust air can be centrally located as long as individual room doors are undercut by a minimum of 38 mm to allow circulation.

- Passive solar gains in south-facing rooms may displace baseboard heating on sunny days. Whole house circulation of solar heat is limited.

- DHW could be supplied by a heat pump system or by an efficient fuel-fired water heater, which could be side-wall vented.

18.4   ADVANCED HOUSES

In 1992, Energy, Mines and Resources Canada sponsored a cross-Canada competition to foster innovation in the housing industry. Houses had to meet energy and environmental technical requirements (the energy budget for each house was 50 percent that of an R-2000 home.)

Demonstration houses were constructed across the country. Each explores a range of new materials, products and systems. Almost all of them employ some form of integrated mechanical system.

Predominant systems included ground source heat pumps supplying both heat and DHW, combined gas space heat and DHW, and integrated space heating and ventilation. As a general rule, where natural gas was available, integrated systems were based on gas technologies; where natural gas was not available, high-efficiency oil-fired systems or Earth Energy systems were employed. The following are some examples:

Nova Scotia EnviroHome

The EnviroHome's heating system takes the best of new oil-fired hot water heating technology and adapts it to the needs of today's low-energy houses. Most of the components are standard items, but the configuration of the whole system is innovative and unique.

A traditional oil system operates for short periods of time and has a low efficiency. The major challenge to adapting an oil-fired system to a low-energy house was to increase efficiency. The project team added an extra hot water storage tank sized to allow the boiler to operate for long enough (10 to 15 minutes) to reach peak operating efficiency. Once the water in this heavily insulated tank is up to temperature, the boiler may be off for several hours. Unlike many standard boilers, the low mass cast iron boiler in the EnviroHome uses no oil during periods when it is not required. Combustion air for the unit is supplied by a separate fresh air duct which is controlled by a motorized damper. The boiler is vented using a powerful sidewall exhaust, eliminating the need for a chimney. (see Figure 18.6)

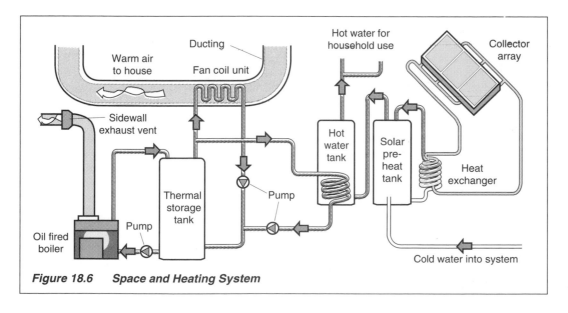

Figure 18.6   Space and Heating System

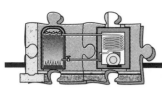

Hot water from the boiler is circulated to two fan coil units (a fan coil unit is similar to a car heater) — one serving the Upper Floor and a larger one serving the Main and Lower Floors — where the hot water is used to warm the air that is then distributed around the house. Using a warm air distribution system rather than standard baseboard heaters offers two big advantages: the same ducting system is used to distribute both heated air and fresh air from the ventilation system as well as move the heat captured by the south-facing glass to other areas of the home. A return air intake located high on the wall above the fireplace helps to prevent hot air from collecting in the peak of the cathedral ceiling.

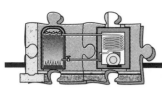

Figure 18.7   *Ottawa Space Heating and Hot Water System*

Ottawa Innova House

Innova House features an integrated heating system that supplies heat for both space heating and hot water from a single, natural gas-fired hot water tank. This direct-vent, hot water heater operates at up to 94% efficiency. The unit is equipped with a water-circulating pump and a heat exchanger coil located in the central air handler that transfers heat from the water to the air, which is then circulated throughout the house.

The water heater tank is insulated with RSI 3.0 to reduce standby heat losses, and the hot water pipes are also insulated.

Water in the tank is maintained at 60°C. However, a tempering valve adds cold water to the hot water prior to distributing it to the sinks, tubs and showers in the house. This process lowers the temperature of the water and saves hot water consumption and energy.

To further conserve energy, a 350 L tank has been installed on the water supply side to the water heater. The tank picks up heat from room air to reduce the amount of energy required to heat the water for household use. In warmer seasons (about six or seven months in Innova House) when heat from the room air is no longer required for space heating, it pre-heats the tank at no charge and reduces the cooling requirements of Innova House. Calculations show that this feature of the integrated heating system reduces energy consumption for water heating by approximately 12%. (see Figure 18.7)

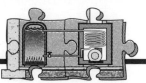

Waterloo Advanced House

The Waterloo, Ontario, Advanced House is heated by a prototype combination of natural gas furnace and heat recovery ventilator. The system uses a mid-efficiency furnace and provides high efficiency (85 percent) at a relatively low cost. The HRV is a regenerative rock bed that recovers heat from ventilation exhaust air and hot furnace gases.

The HRV is split into two beds that operate independently; one heating intake air, the other removing heat from the exhaust stream. When one bed is heated and the other is cooled, the air flows through the beds are reversed. Using the HRV to capture furnace heat means that acids in the condensate are handled by the rock bed. As a result, the furnace does not require an expensive stainless steel heat exchanger. (see Figure 18.8)

P.E.I. House

In this house, heating, cooling and some DHW needs are met by a prototype ground source heat pump, the Chinook Phi Beta. Typically GSHPs are linked to a forced-air heating system; in this case, the GSHP is linked to a hydronic (radiant floor) heating system. In addition, the pipe containing the circulating fluid which extracts heat from the ground is a revolutionary "spiral loop" developed by the National Research Council. The spiral allows greater pipe lengths to be installed in shorter trenches, substantially reducing installation costs — a COP of 3.6 has been measured.

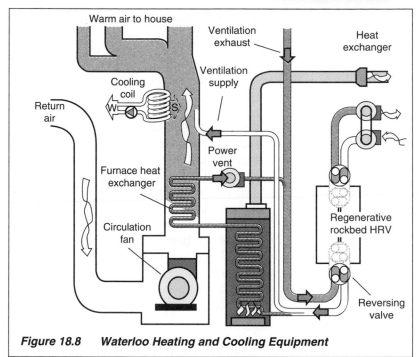

Figure 18.8   Waterloo Heating and Cooling Equipment

Maison Performante

Located in Laval, Quebec, this project combines passive solar heating with a ground source heat pump. Approximately 35 percent of space heating needs and all DHW needs will be realized through a large sun space and roof-mounted solar collectors. Excess heat energy is collected and stored in a 10,000 litre cistern located under the garage. That "energy" is then extracted by a ground source heat pump. The concept is to keep the cistern as cool as possible, without freezing the water, to enhance heat transfer. When the water in the cistern falls to 4°C, the GSHP reverts to its more conventional mode, extracting heat directly from the ground. As a further aid to reducing purchased energy costs, the GSHP transfers heat in the summer from the house to the ground without using a compressor.

Saskatchewan Advanced House

Using sixty highly efficient sealed glass solar collection tubes and employing both natural gas and electric domestic hot water heaters, the solar collection system of the Advanced house aims to capture and store enough heat to supply 40% of the space heating and 80% of the domestic hot water requirements for this house.

Solar heat is transferred from the rooftop collectors, via an ethylene glycol loop circulated by a photovoltaic pump, to copper heat exchange coils in a storage tank. Heat from the storage tank is then transferred by Thermosiphon lines to both water heaters. Both domestic water heaters are coupled by a heat exchanger to the radiant floor heating system for the house, and can be switched on to supply back-up heat. When hot water is needed for normal household activities, it is drawn equally from both tanks.

The Advanced House uses a Hydronic Radiant Floor Heating System which consists of 3,600 feet of cross-linked polyethylene distribution tubing which circulates warm water to radiate heat from all the floors of the house. The tubing is embedded in a layer of concrete, which in turn radiates heat evenly and consistently throughout each room. Each room may have its own controls, so heat can be directed to the areas that require it.

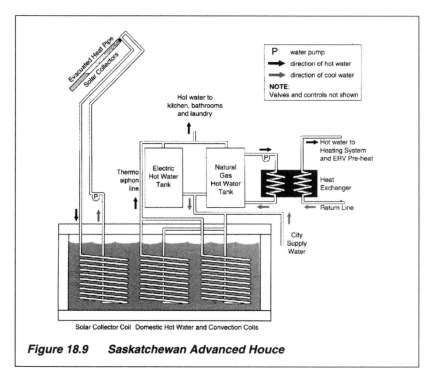

Figure 18.9   Saskatchewan Advanced Houce

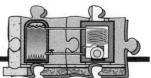

Brampton

The original Advanced House, built in Brampton, Ontario, in 1989, features a prototype IMS—*The Solmate*. It combines space heating, water heating, passive solar storage, ventilation, heat recovery, and partial cooling. In the summer, the system replaces all of the house's mechanical equipment dealing with thermal energy, with the exception of the refrigerator (see Figure 18.10).

In addition to providing standard mechanical functions such as space heating, the system also provides heat recovery from grey water and heat storage. The addition of grey water heat recovery to the heat recovery of ventilation air enables the system to make use of all waste energy from the home. With the addition of storage, the system is able to make use of passive solar heat gain and internal gains in an efficient manner.

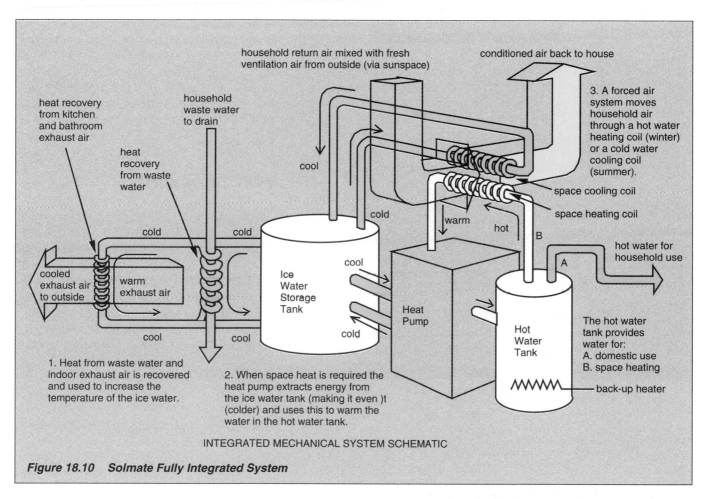

Figure 18.10 Solmate Fully Integrated System

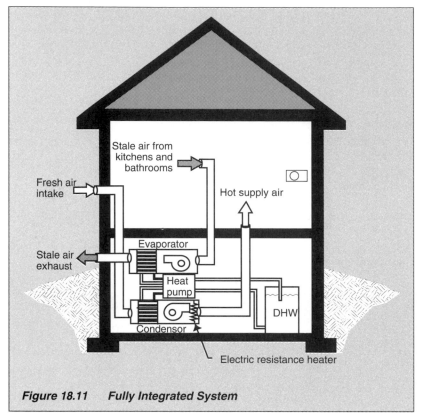

Stale air from
kitchens and
bathrooms

Fresh air
intake

Hot supply air

Evaporator

Stale air
exhaust

Heat
pump

DHW

Condensor

Electric resistance heater

Figure 18.11   Fully Integrated System

18.5   FULLY INTEGRATED SYSTEM

Figure 18.11 shows how full integration can be achieved using heat recovery from heat pumps for space heat, DHW and ventilation. The heat pump HRV supplies heat to the hot water tank providing domestic hot water and space heating. Space heating is accomplished by circulating hot water from the storage tank through the fan coil unit of the forced-air distribution system.

All systems are interactive. Cold fresh air is mixed with warm return air then passive solar gain before being circulated through the house. Exhaust air is used to provide some space heating and domestic hot water. Auxiliary electric coils provide backup when demand for heat exceeds the exhaust air capacity.

18.6   CONTROLS

Almost all of the Advanced Houses utilize some form of home automation for efficient energy use. Through the home automation system, each of the home's major appliances and mechanical systems communicates with other energy users in the house. An automated system, for example, is able to take advantage of "free" air cooling by calculating the cooling potential available within the house and making the necessary adjustments in the mechanical system's operation.

Automated systems will likely become a major component in the efficient operation of these new technologies. (See Chapter 20 for an expanded discussion of home automation systems.)

Part *4*

EMERGING TRENDS

Chapter **19**

ENVIRONMENTALLY SUSTAINABLE HOUSING

Since the mid-1970s and through most of the 1980s, most innovations in house construction dealt with making the envelope and mechanical equipment more energy efficient. More recently, environmental sustainability has become a major focus of housing technology research and development. Many of these advances are being considered as part of the new R-2000 Program Technical Requirements. For the home buyer, many green building practices result in lower house operation and maintenance costs. For the home builder, these practices represent a marketing edge in an era where people are increasingly concerned about the state of the environment.

This chapter looks at some of the technologies and processes involved in designing and building to higher levels of energy and resource efficiency. It also addresses some of the issues surrounding marketing "green".

Readers may also want to acquire CHBA's publication "Environmental Choices for Home Builders and Renovators".

19.1 ENERGY EFFICIENT LIGHTING

The typical incandescent bulb found in most Canadian homes converts only 15 percent of its electricity to light energy — the other 85 percent is lost as wasteful and expensive heat. In an energy efficient home, the lighting portion of purchased energy can amount to as much as 10 percent.

By replacing conventional lighting systems with energy efficient lighting, savings of 800 to 1,000 kWh per year can be achieved. Although more expensive than conventional incandescent fixtures, energy efficient lighting will pay for itself through savings on the power bill.

This section looks at different options for energy-efficient lighting.

19.1.1 Lighting Design

Good lighting design takes into account both natural and artificial light. Using sunlight wherever possible helps reduce the need for artificial light and reduces a home's demand for purchased energy.

The typical residence will have lights with many different burn-time durations or "duty cycles" (the proportion of time a light is on per day, expressed as a percentage). Where lights are placed, and how many are used, will play a big role in energy consumption.

Some basic rules should be followed in the design of effective systems.

- Don't use high wattage bulbs where low wattage bulbs will do.

- Provide a mix between general room and task lighting.

- Don't put too many lights on one switch — especially with track lighting — but divide them into logical groups and put them on individual switches so areas can be lighted independently.

- Wire each room's switch box close to the door or stairwell for ease in turning off the lights when leaving a room.

Task lighting is designed for a specific task that must be done under particular lighting conditions. For example while one family member is watching television, another may want to read. Task lighting using downlights or a reading lamp in one area of the room will allow someone to read without interfering with the television watcher. An overhead light might produce glare on the screen.

19.1.2 Lamps and Fixtures

Check with your suppliers for the latest energy efficient lighting products and their suggestions for lighting design. Use them as a resource; they often have trained lighting consultants on staff to help with such questions. You'll need to determine which lamps can be used with which fixtures; all are not mutually compatible.

Fluorescent Lighting

Fluorescent lighting saves 80 percent of energy compared with incandescent lighting. Energy efficient fluorescent lamps (compact and standard) can be justified in many locations in the home.

Hard-wired fluorescent fixtures are encouraged for new building in many locations. Compact fluorescent lamps can easily be used in all bare bulb, non-decorative sockets (see Figure 19.1a). "Daylight" and "warm light" options overcome the blue glare associated with traditional "cool blue" fluorescent tubes — but at a cost in energy efficiency.

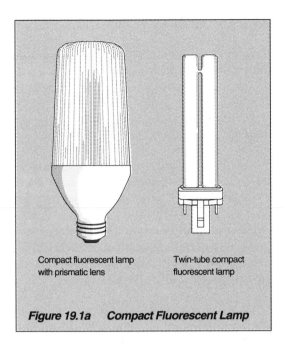

Compact fluorescent lamp with prismatic lens

Twin-tube compact fluorescent lamp

Figure 19.1a   Compact Fluorescent Lamp

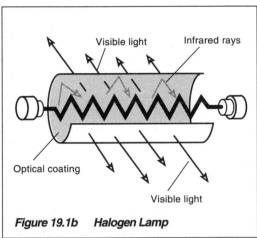

Visible light   Infrared rays

Optical coating

Visible light

Figure 19.1b   Halogen Lamp

Compact Fluorescents

A compact fluorescent lamp can cost 16 times a regular incandescent light bulb, but it lasts ten times longer, requires 75 percent less energy, and more than pays for itself in energy savings. With a separable ballast, they are rated to last about 45,000 hours.

The lighting design should be discussed with the buyer. Installing a ballast, for example, removes the owner's option of replacing compact fluorescents with regular incandescents or halogens.

The biggest barrier to consumer acceptance of compact fluorescent is price. Other drawbacks include their odd sizes and shapes, and the fact that they cannot replace higher-wattage incandescents, or be used with dimmer switches.

Halogen

Halogen lamps (see Figure 19.1b) provide a more intense light than a standard incandescent, which some people find more pleasant. If used at their full light output, they use about 15 percent less electricity than incandescent lamps. But fluorescent lamps are still more energy efficient.

Most halogen lamps are also dimmable but dimming inhibits the halogen cycle, the basis for its longer life and higher efficiency. The lamp's lifetime will be shorter than specified by the manufacturer, negating the efficiency improvement.

Halogen lamps are often purchased in large sizes: 300 watts is very common — more light than needed for most residential applications. Instead of dimming, if you need less light and still want halogen, buy a lower powered unit to begin with.

Halogen is an excellent choice where an intense beam of focused light is needed, such as in downlighting, spotlighting or for creating a "wash" of light in torchieres. Particularly good for exterior usage, halogen lamps are also good inside where a particular atmosphere is desired.

Into the Future

Lighting technology has come a long way since the first incandescent bulb experiments. New to the residential market is an electronic light bulb or an induction lamp (see Figure 19.1c). The technology uses high-frequency radio waves to generate light in a manner similar to fluorescents, but the lamp has no filaments.

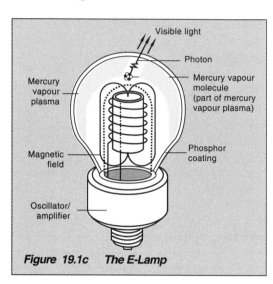

Figure 19.1c   The E-Lamp

19.2   APPLIANCES

Energy efficiency in appliances is one area in which the builder's influence can be felt. Over 250,000 appliances are purchased by the housing industry every year, appliances which could, over the course of their lifetime, cost $300 million to operate.

Six major appliances — the fridge, freezer, dish washer, range, clothes washer and dryer — account for 14 percent of total energy use in the home. By installing energy efficient appliances, builders can cut appliance electrical use in half. The electricity saved by buying energy efficient appliances can add up to thousands of dollars.

The energy consumption of an appliance will depend on the model selected and the frequency of its use, but savings to the homeowner can approach 20 to 60 percent of the energy average appliances would use.

The Energuide Directory lists the major appliances sold in Canada and includes their Energuide ratings (see Figure 19.2). The lower the number, the more efficient the appliance. The directory is available free from Natural Resources Canada in Ottawa.

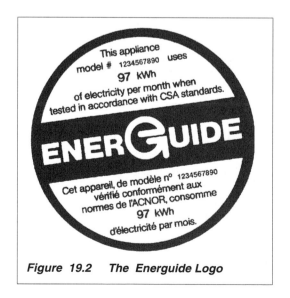

Figure 19.2   The Energuide Logo

19.2.1 Selecting Appliances

Refrigerators

One of the biggest problems with refrigeration is the dangers of CFCs represent to our environment, particularly their impact on the ozone layer. The challenge of reducing or replacing CFCs continues to preoccupy manufacturers.

In the meantime, further improvements are being made in energy efficiency. Manufacturers are extending the technique of super-insulated chest freezers to refrigerators and refrigerator/freezer combinations.

The energy efficient refrigerator now averages about 900 kWh/year compared with the average 1,250 kWh/year and 1,500 to 2,000 kWh/year for those aged to about 15 years. Even greater efficiencies will result when products currently being developed are released. Models in development are so efficient they can be powered by photovoltaic solar panels.

The energy consumption of refrigerators varies greatly. Features such as ice makers, through-the-door ice makers, automatic defrost, and top or side freezers will give different energy costs depending on the feature.

Where the unit is located (in an air-conditioned or non-air-conditioned kitchen, next to the stove, in the garage, in direct sunlight), the number of times the door gets opened (usually this increases with the number of children in the home), and food loading (how often and at what temperature) will all contribute to the annual energy use for individual refrigerators, which can often differ by as much as 40 percent from their test ratings.

Energy consumption also varies over the year. As with heating and ventilation, refrigerators have a large seasonal component to their energy use.

Freezers

In the 1980's, a new generation of super-insulated chest freezers was developed. With only a slightly reduced interior volume and wall insulation as thick as 10 cm (about 4 in.), the new units offered energy savings of up to 40 percent and up to 65 hours of cold in the event of a power failure.

Again, the model choice — even within the same price range — can greatly affect energy costs. Automatic defrost models are particularly high energy guzzlers. An anti-sweat heater on the freezer located in a dry area can be switched off for savings of 120 kWh/yr.

In future, look for a freezer system capable of reducing energy consumption by cooling only those interior "zones" areas which contain food.

Stoves, Ranges and Microwaves

The stove to look for is a well-insulated combination conventional and convection oven — convection cooking is the more efficient of the two because it cooks faster.

A cook top that includes halogen elements is more efficient and safer; heat stays where it is needed, under the cooking pot.

A microwave oven is the most energy efficient way to cook, unless it is used to thaw frozen foods; then it uses more energy than a regular oven.

Dishwashers

Compare when looking for a dishwasher to find one of the most efficient on the market. Look for features like power saver cycles, choices in water temperature and wash time, and water-saving features.

Clothes Washers

Front loading, European style washers are usually 35 percent more energy efficient than top loading units and they save water too. The horizontal axis variety clothes washer needs to be only partially filled with water, and generally consumes two-thirds less energy and one-third less water than the North American vertical axis units.

The latest developments include a higher spin speed in the clothes washer: up to 1,600 revolutions per minute. This extracts more water from the clothes before they are put into a dryer in a process seventy times more energy efficient than the thermal extraction technology employed in tumble dryers.

Other features to look for include washers that heat their own water, wash at a lower temperature, use weight and fabric recognition systems, and offer even more energy savings.

Dryers

New heat pump and microwave dryers should be available in the mid 1990's. These technologies will save 50 percent or more of clothes drying energy.

19.3   WATER CONSERVATION

Working in a water-rich nation with lots of room for sewage, Canadian builders have not traditionally designed for water conservation. Canadian water prices have been some of the lowest in the world. Residential water use is extremely high. Canadian homes are second only to those in the United States.

Increasingly, the ability of our supplies and our existing patterns of water infrastructure to support new development is becoming uncertain. More than 25 percent of Canadian water supply systems are over 50 years old. Some regions, dependent on ground water, face crises servicing even their existing communities in the long term. New developments are being put on hold in some areas of the country. In other areas, domestic water and sewer charges are increasing substantially.

Builders and developers across the country are now installing products and systems which can dramatically reduce water consumption. Transfer payments from senior levels of government are diminishing, so municipalities are finding infrastructure extensions and maintenance prohibitively expensive. Water conservation is increasingly a requirement for development approvals.

Beyond the requirements, legislation, and housing market demands for water conservation, it makes good environmental sense. Incorporating these features into the homes you build sets a marketable example. They can be low cost demonstrations that your company is pro-actively designing for an increasingly environment minded buyer.

Household water consumption can be minimized just by making slight design changes outside and indoors. Improving conservation inside the home often requires little more than specifying alternative, more efficient products. This section outlines the major indoor water users in order of the relative savings possible: first toilets, then showers, faucets and washing machines.

Outdoor water conservation initiatives might incorporate irrigation, landscaping methods and ways to reuse rainfall.

19.3.1 Designing Homes for Indoor Water Conservation

At least 50 percent of a typical household's water bill comes from interior uses — and fully 95 percent of that amount is used in bathrooms and laundry rooms. The total figure can be substantially reduced for relatively little cost. The R-2000 Program now incorporates requirements for water conserving appliances.

Water-saving fixtures now cost about the same as regular "water-guzzlers", and work just as well. Fixtures with consumption rates similar to European models are becoming available here.

Incorporating water-saving fixtures in houses makes sense. Each person in a Canadian household uses an average of 350 litres of water every day. Efficient appliances can save the family hundreds of dollars each year at current prices. With water and sewer rates escalating, the savings will be even greater in the future.

19.3.2 Water-Efficient Toilets

Almost half of the average household bill is spent just to flush the toilets.

Moderately efficient and water-saving toilets can have a substantial impact on annual water costs (see Figure 19.3). After more than a decade of design innovations, their performance in the laboratory and in the field is well-documented. Although cost vary considerably from fixture to fixture, there is no little correlation between price and performance.

Low-flow models using 6 litre (1.3 imp. gal.) or less per flush are available from major manufacturers for close to $200. Some low flush toilets are sophisticated, using compressed air or water pressure on special diaphragm chambers to provide a forceful flush with little water. Others use a more conical bowl and redesigned water channels to enhance performance.

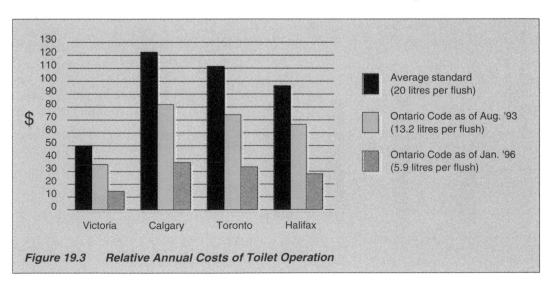

Figure 19.3   Relative Annual Costs of Toilet Operation

Since a toilet should last for decades, parts availability is crucial. The fewer specialized mechanisms and moving parts, the better. Installing a warranted unit from a known, reliable manufacturer should reduce future call-back problems.

There is virtually no "consumer cost" associated with installing water-saving models. Many look and perform so much like "regular" toilets that users cannot perceive a difference, except in their utility bills.

19.3.3 Water-Efficient Shower Heads

Nearly one-third of indoor water is used for bathing. Low-flow shower heads can reduce both water and energy costs by at least 50 percent when compared with standard fixtures (see Figure 19.4).

To use less water, some shower heads restrict the water flow by pulsing it. Others mix the water with air. Both systems give the user the impression of a more powerful flow and are reported to perform well.

Again, there is no general correlation between price and reported performance. Some shower heads selling for more than $20 received no better reviews than models that cost under $10. Constructed of plastic, metal, or combinations of both materials, these fixtures are available at most plumbing supply outlets in a range of styles and colours.

Low-flow shower heads have met with widely varying reactions from homeowners. However, independent testing has shown no relationship between flow rates and perceived performance. Some low-flow units were deemed to deliver a better shower than their high-flow counterparts.

19.3.4 Low-flow Faucets

Typical bathroom and kitchen faucets have flow rates that average 13.5 litres (3 imp. gal.) of water per minute. That is far in excess of what's usually needed or can be used. Aerating faucets give the illusion of more water flowing from the faucet than is actually the case. For washing hands and brushing teeth a flow rate of 2 litres per minute is quite acceptable. In kitchen sinks and laundry tubs, flow rates of 6 to 9 litres per minute are often sufficient.

Some faucets are spring loaded and turn themselves off either immediately or after a brief delay. These have been marketed mostly for commercial applications but are inexpensive and may be a practical way to minimize water waste in households.

Higher tech options, such as infra-red sensor mechanisms built into taps, have been employed to automatically cut the water flow when hands are removed from beneath the tap.

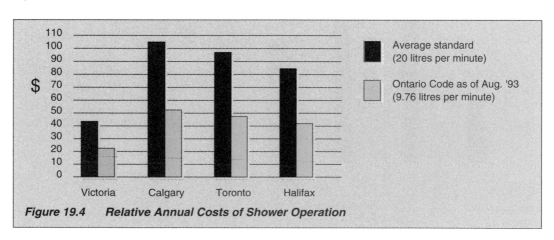

Figure 19.4   Relative Annual Costs of Shower Operation

19.3.5   Water Saving Appliances

Dish- and clothes washers are being designed to function with less water. Conventional washers can average more than 1,200 litres (266 imp. gal.) per month, depending on household size and frequency of use.

Front loading washing machines, which tumble the clothes through the water rather than immersing them entirely, use up to one-third less water than the top loading machines more common in North America.

19.3.6   Outdoor Water Use

Irrigation water for lawns and gardens can account for up to 80% of overall domestic demand in the spring and summer. So much water is used that ,municipal supplies can be strained in peak demand periods. As the peak demand is the determining factor in planning for new developments and housing, outdoor water use can have far-reaching impacts.

There is a real need for improvement. Between one-third to one-half of this water is wasted: never reaching the plants for which it was intended, but rather being misdirected or evaporated in warmer weather.

To overcome these problems, a drip irrigation system can be installed prior to sodding or seeding. These systems, which are really just soaker hoses laid under the turf, supply water directly to the root zone of the plants.

Automatic watering systems further increase the efficiency of all irrigation. Outfit hose couplings with timers to ensure that watering is done by night, when demand on the municipal water system and evaporation losses will be lowest.

Moisture sensors can make the system even more "smart" by watering the lawn only when required and shutting off programmed watering during periods of rain.

Together, these measures can cut total household water consumption during the summer months by up to 50 percent.

19.3.7   Landscaping for Water Conservation

Traditional turf grass requires a great deal of irrigation. A typical suburban lawn of about 350 square metres needs as much as 200,000 litres of water through each growing season.

Lawn areas are increasingly being seen as long-term liabilities. Smaller households, especially are not interested in spending a great deal of time on the upkeep of their lawns. Alternative landscapes provide opportunities for significant and permanent reductions in household outdoor water use.

Reducing the size of turf grass areas, concentrating them in back yards, and surrounding them with native plants (which have greater resistance to insects and diseases), means far less water is used for irrigation. It also means reduced demand for fertilizers, pesticides and herbicides.

Wooden decks, gazebos, interlocking stone areas, walkways and boxed-in planting areas or gardens can preserve the earthy, rustic character of landscapes while saving the homeowner substantial maintenance time.

Top soil stripped from construction sites should always be replaced. A deep layer of top soil creates an absorbent base to retain water where grass needs it most.

19.3.8 Cisterns

As simple as a barrel, but more commonly an in ground or basement tank, cisterns capture rainfall from rooftops and other hard surfaces. Often in abundant supply, this soft water lacks the mineral salts common to ground water and is ideal for irrigation. Hundreds of cisterns in Nova Scotia are providing substantial percentages of monthly water requirements — especially in remote areas.

As much as 90 percent of the annual rainfall on the roof can be collected, depending on the degree of its slope, and every few centimetres of rainfall provides approximately 20 litres of water per square metre.

The cistern is best constructed of high density concrete, vibrated as it is cast in place, and allowed to wet cure before being put to use. A tight, opaque cover should avoid problems with mosquitoes or algae.

Water storage in a cistern also reduces the flooding impact of storms, lowering urban runoff. Cistern water storage can be an integral part of a storm water management plan.

19.3.9 Conclusion

Water efficient design saves far more than it costs. Low-flow shower heads and toilets, for example, cost the builder no more than standard fixtures that waste water.

Water efficient homes simply cost less to operate. At 1992 rates, a Toronto family of four will spend about $5,000 on water over 20 years of home ownership. The same four people in Calgary will spend over $13,000.

Water-efficient toilets, shower heads, and supplying drip irrigation hoses for lawn watering would add less than $200 to a house price with a payback of less than three years at Toronto prices and closer to one year in Calgary.

Building for the future shows your foresight. Homes that offer less expensive long-term operations and which don't compromise the quality of living that homeowners have come to expect are both environmentally sensible and easier to sell.

19.4   CONSTRUCTION WASTE MANAGEMENT

Wastes produced during site clearance and construction of new homes and residential renovations account for more than five percent of the total volume of wastes currently taken to landfill sites in Canada.

Bans and restrictions on a number of construction wastes, hazardous, and recyclable materials are already in place in many parts of Canada, and more are being considered all the time.

Cutting back on the amount of wastes slated for landfill can help builders:

- reduce the cost of materials purchased
- reduce haulage and tipping fees
- improve efficiency through altered construction practices
- improve the energy efficiency of the building envelope by reducing the amount and types of materials through which heat loss can occur, and
- improve the company's public image by demonstrating a responsibility towards the environment.

The Canada Mortgage and Housing Corporation publication *Making a Molehill out of a Mountain II: Implementing the Three R's in Residential Construction* suggests that builders review their operations to build a waste reduction plan around the three R's:

- reducing waste at source
- reusing what would normally be landfilled, and
- recycling materials for which there is no immediate reuse.

19.4.1   The Three R's Audit

Construction wastes vary across the country according to local practices and materials used. Figure 19.5 illustrates typical wastes generated in Ontario projects.

Wastes will vary according to your design, construction techniques, site management procedures and materials. An audit will tell you what wastes you are producing and what they are costing your company. Identifying conventional waste volumes, determining where wastage can be reduced, and identifying what materials can be reused and recycled is the foundation of a waste management plan.

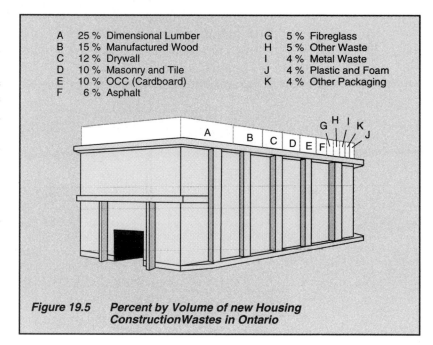

A	25 %	Dimensional Lumber
B	15 %	Manufactured Wood
C	12 %	Drywall
D	10 %	Masonry and Tile
E	10 %	OCC (Cardboard)
F	6 %	Asphalt
G	5 %	Fibreglass
H	5 %	Other Waste
I	4 %	Metal Waste
J	4 %	Plastic and Foam
K	4 %	Other Packaging

Figure 19.5   Percent by Volume of new Housing ConstructionWastes in Ontario

A small builder may only need to examine wastes resulting from one or two houses. A large builder may need to track wastes for a period of months to obtain a complete picture. An comprehensive audit system should include:

- Track the waste contents of each bin over a specified amount of time.

- Use a new audit sheet each time a bin is emptied.

- If you are source separating materials already, one chart can be used to track several bins, but for one collection period only.

- As much as possible, keep track of the different construction phases by starting a new chart as a new phase begins. This step will help you zero in on those operations that produce the most waste.

- Checklists will vary according to the type of work each company performs.

A good understanding of your current waste profile will allow you to develop your plan to implement the three R's — reduce, reuse and recycle.

19.4.2   Reduce

Reducing waste at source (see Figure 19.6) is the most efficient and effective of the three R's. Less waste means less materials purchased, and fewer materials to dispose of, with direct savings to the company.

You can reduce wastes in a variety of ways, by using materials more efficiently or through smart purchasing and handling practices, including:

- altering floor plans at the design stage to eliminate redundant materials.

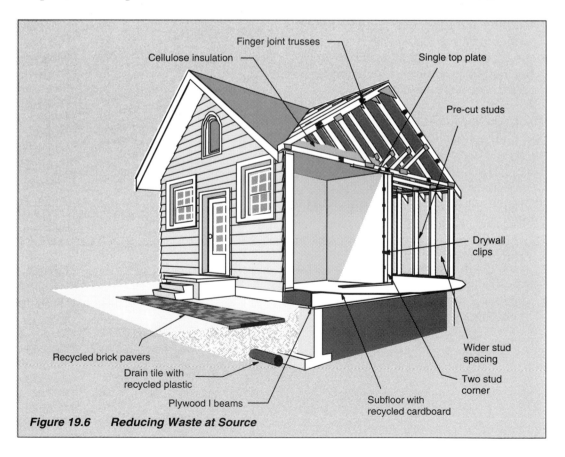

Finger joint trusses

Cellulose insulation

Single top plate

Pre-cut studs

Drywall clips

Recycled brick pavers

Drain tile with recycled plastic

Plywood I beams

Subfloor with recycled cardboard

Wider stud spacing

Two stud corner

Figure 19.6   Reducing Waste at Source

- leaving trees standing where possible or chipping roots and branches on-site at the site-clearing stage.
- improving materials storage procedures during construction to minimize weather related damage.
- reducing the amount of packaging left at the site by suppliers.
- purchasing materials such as fasteners, paint, caulking and drywall mud in bulk containers
- buying more durable materials such as kiln-dried wood to reduce the amount of material rejected due to warping and shrinkage
- buying materials either entirely or partially made of recycled materials — using recycled materials reduces the amount of virgin material harvested and consequently reduces the amount of virgin material that ends up as landfill
- making your sub trades responsible for dealing with their own wastes, or at least requiring subs to place their wastes in your separated bins, and
- developing a centralized cutting and storage area on a multi-unit job to keep all wood waste in one spot — this system promotes reuse of off-cuts, allows for better collection of waste wood, and makes clean-up a faster operation.

Manufacturers are already offering moving into less wasteful products and products with less packaging. The number of recycled construction materials is growing at a tremendous rate. Builders can help this process by demanding resource conserving products.

19.4.3 Reuse

Reusing what would normally be landfilled is the second most important of the three R's. All too often, perfectly useful materials are thrown into the disposal bin on the basis of ease. The following area few of the ways that builders and renovators can save money and increase quality by reusing "waste" materials.

- Off-cuts from framing lumber can be reused as bridging, blocking or forming stakes.

- Excess insulation from exterior walls can be reused as sound barrier material in interior walls — using excess in this way is cheaper than paying for its haulage.

- Drywall pails can be used as tool caddies.

- Plastic packaging can be reused as garbage bags or as protective covers for materials.

Builders can also pool waste resources, storing them at a common location, from where they can be diverted to different jobs or sold for profit. A number of centres already divert construction wastes in reusable form.

19.4.4   Recycle

The recycling logo (see Figure 19.8) is a familiar sight, but recycling is the last option that builders should consider. Any production process that makes new materials out of old ones creates new by-products that in turn become waste. Recycling is almost always superior to throwing usable material away—anything left in the dumpster should be considered for recycling.

Materials already being recycled in various parts of the country include asphalt, wood, drywall, cardboard, masonry, metals, glass, plastics and paints. Plastics, for example, are being recycled into numerous products like drainage tiles, sump liners and wood substitutes.

Some wastes can be sold to recyclers. If recyclers charge to remove waste, the fees are generally less than landfill site tipping fees. In some cases, if amounts warrant, the recycler will provide bins and haul the waste free of charge. However, most builders will set up their own bins for source separation. Wastes such as wood, drywall, cardboard, etc. must be separated and stored independently.

The economics of recycling are much more attractive when the material is "clean" and not contaminated with another form of waste. Where there is a market, clean waste always brings a higher price.

Bins for each kind of waste will have to be clearly labelled, identifying what materials go where. On smaller jobs, a single bin with a series of dividers can be used to keep wastes separate from each other.

To keep your waste clean, consider locking it up—someone else's garbage bag inside an otherwise clean load can spoil the entire load. Larger companies might consider setting up a blue box system for residents who move in early and do not have garbage pick up. As your bins fill with household wastes, you realize for a period of time you are in the garbage business.

The list of potential sources and uses of recycled materials grows daily. More specialized equipment for recycling and more specialized recycled materials will begin to appear. Find out what materials can be recycled locally, and be aware of new developments as new uses for old materials become more practical and widespread.

19.4.5   Summary

The common practice of leaving mixed garbage in random piles around the site must be discontinued if source separation is to work.

The introduction of new practices may require some planning. However, once these changes have been properly implemented, they will become a regular component of site management procedures. Often, it is simply a matter of substituting one routine for another.

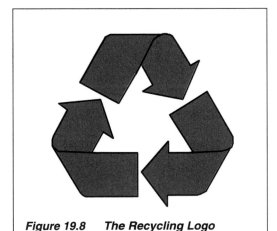

Figure 19.8   The Recycling Logo

19.5 GREEN MARKETING

If you are going to build environmentally sustainable housing, then you have to let people know. To develop your market niche, you must highlight the difference between your houses and others in the same market.

In other sectors, companies are leading consumers, selling them on the benefits of environmentally responsible purchasing. In the building industry, you too have to let most people know what is available. As customers know more, they are asking questions. You have to be aware of new trends and developments so that you can answer their questions and fill their requirements.

Money isn't always the big issue for this kind of client; it's often a health or a "from the heart" issue. Selling to the environmentally aware will give you that "in" over another builder.

If you plan to market yourself as an R-2000/green builder, the following are some basic rules to consider when developing your marketing strategy and implementing your green marketing plan.

19.5.1 The Golden Rules

Know your market.

Just because a lot of people are environmentally aware doesn't mean that they will buy your "green" houses. Market research is necessary to help you identify your best target market (by age, household size, income level, etc.) and the product which they are most likely to purchase.

If you are just starting, take time to develop a company-wide approach. Define your approach to environmentally responsible housing, and clarify it to your target market.

Don't do it if you don't mean it.

The whole business has to be on side. If you are going to sell yourself as an environmentally aware builder, those values must be reflected in the physical office and the way your company does business. All people who are working with you must be aware of and "buy in" to your company's objectives.

If you are going to do it, make sure you know what you are doing.

You may have to do a lot of research to find out about acceptable products and methods. The knowledge will be used later in promotional materials and advertising, and in your marketing strategy. You may also find new and simpler ways of doing things.

Many examples of new technologies and products are out there. Go with products that are more readily available. Absorb new technology into your business practices and procedures as they become proven.

Some companies are making their employees the company authorities and experts. If you can free up staff, send them on seminars or give them time to do the needed research. Staff retreats can provide training that spreads your philosophy through the organization. Bring in outside help as required.

Beware. Don't call yourself a green builder.

Don't put yourself up on a pedestal — people will be watching, some even hoping you will fall. Even if you are being environmentally responsible in all ways, shapes, and forms, be careful in your promotions.

Green marketing considerations range from all green practices covered in this manual to your marketing materials and then some. Don't flood the market with junk mail or print your brochure on non-recyclable, glossy stock, for example!

Be explicit.

When you are developing marketing materials, say exactly what you are doing on specific projects and what new systems, methods, or products are being utilized. Sales personnel must also be aware of what makes you different from the next builder or renovator.

Tell people exactly how you are solving environmental problems. By bringing the community on side, one developer shaved two years off the rezoning process — two years closer to selling, with two years' less taxes to pay and interest to carry!

Tell people where you are introducing new practices. For example, if you use composting as part of a kitchen package, consider selling the home as a better thought-out product.

Educate your salespeople.

Educating your salespeople to sell environmental benefits to prospective purchasers is the key to selling green. Provide the information they will need ahead of time. You have to make sure they understand what you as the builder are trying to achieve.

Your salespeople must be able to recognize which of your customers are vitally interested in environmental features and the benefits associated with this approach. Similarly your salespeople must understand when the "green" approach is not an important part of the sales process.

Choose your suppliers and sub trades well.

Let your suppliers and subs do some of the footwork or supply some research for you.

Many subs are already using recycled materials, and many products coming through suppliers already have recycled content or revamped processing procedures. If they are working towards your company objectives, you may as well market the fact.

Chapter *20*

HOUSING INTO THE FUTURE

Over the next decade, there is likely to be greater change in the housing industry than ever seen before. Not only are energy efficiency and environmental responsibility, as discussed in the previous 19 chapters, likely to gain widespread acceptance, but the face of Canadian housing will also be altered by technological and societal change. Successful builders in the 1990s will be looking to change design and construction practices to enhance affordability, to improve quality, and to increase their market share by offering more choices to Canadian home buyers. Key changes in society that Canadian builders will need to adapt to include:

* an aging population;
    - the oldest baby boomers are nearing 50 — the youngest have already turned 30
    - one in nine Canadians is 65 or over, and the percentage is expected to grow
* an increasing number of people with physical handicaps;
* the need for affordable housing for a broader segment of the Canadian public; and
* the increasing use of technology in manufacturing and within the home.

This chapter takes a look at many of these trends. It is by no means comprehensive; rather it provides an overview of several of the key issues which are likely to affect the design and construction of Canadian homes in the future. It is hoped that the ingenuity of the building industry will continue to develop innovative approaches to many of these challenges.

20.1 HOUSING THE AGED AND DISABLED

One in every nine people in Canada today is aged 65 or over. And the Canadian population is aging (see Figure 20.1). By the year 2030, an estimated one in four Canadians will be over 65 years of age. As part of the aging process, physical disabilities will become more commonplace. As many as 3.3 million Canadians, or 13 percent of our population, have some form of activity limitation. Multiple disabilities are more common among seniors, and the likelihood of having disabilities increases with age.

People with disabilities and the elderly are vocalizing that they no longer wish to be separated from their community. And as a society we are recognizing the need to better meet the needs of the physically challenged members of our communities. To contribute to society as full working citizens and to be part of the community at large, they require housing options that will enable them to maintain or regain independent lifestyles.

Because much of the Canadian housing stock has not been designed to accommodate the needs of people with disabilities, even basic activities such as entering the home, moving from one floor to another, preparing meals, using the bathroom or operating kitchen appliances are performed with great difficulty.

Homes constructed today are the houses many Canadians will be living in as they age. Stairs, kitchens, bathrooms, bedrooms, storage space, windows, and entrances will all require modifications for our aging population. Many of these modifications — grab bars in bathrooms, accessible bathtubs and showers, continuous hand railings on stairs, ramps or stair lifts, main-floor bedrooms, accessible kitchen cupboards and counters, etc. — are much less expensive when allowed for at the design stage than when required in a retrofit.

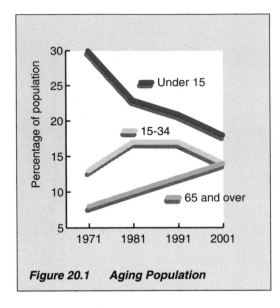

Figure 20.1   Aging Population

20.1.1   Design Principles

Building for the aging or disabled often requires only minor modification to current designs. The ground floor of a two-storey building should contain a living room, a kitchen, at least one bedroom, and a complete bathroom that is fully accessible to a person using a manually operated wheelchair. Occupants must also be able to access the main entrance of the house from the parking area.

Safety and independence through accessibility are key points. Garages can be modified to allow for wheelchair unloading and to provide sheltered entrance to the house. Wheelchair accessible ramps can be integrated into conventional house design. Entrance doors can be upgraded to include wireless entry systems and lowered door viewers. Security systems can be installed with built-in speakers and a siren in case of fire, break-in or other emergency. A wheelchair lift in the staircase can provide access to other floors, and a continuous smooth handrail can provide support and guidance.

Windows should be sized large enough and located low enough to view the outside from both a seated and standing position. Motorized window controls and casement windows with large handles can facilitate operation. Sliding patio doors can provide easy access to yards. Some skylights may be equipped with a special pole enabling them to be easily opened from a wheelchair.

Slightly wider hallways and doorways and a well-planned, open floor plan should allow access to wheelchairs in all rooms. Installing levered door handles, rocker light switches and electrical outlets at locations and heights accessible to everyone are simple design modifications. Installing cupboards with full extension drawers and shelves, lower upper cabinets, and adjustable rod and shelf heights in closets, and providing knee space under sinks, lowered counters and workspaces, will better accommodate the needs of individuals in wheelchairs.

A good accessible kitchen floor plan (see Figure 20.2) will avoid through-traffic in the work area. Countertops and the cook top area can be lowered to allow both wheelchair and standing access to work areas. Cabinet doors can have side-hinged, sliding and flip-up options, and may incorporate valence lighting beneath. As well, a number of options are available in appliances, depending on occupant preference. Look for front-mounted controls.

The entire bathroom should be designed as a "wet" area. The floor of the shower area should be continuous with the rest of the bathroom floor, eliminating thresholds. Families with small children or pets that need bathing can also benefit from the accessible, safer, more functional design.

20.1.2 Issues for the Visually and Hearing Impaired

Various technological options are becoming available for people with low vision and hearing disabilities.

For home safety, solid risers and treads should be painted or carpeted in contrasting colours. Visual/tactile warning strips on nosings and landings make stairs more visible. Non-glare flooring and contrasting colours throughout a home make it more accessible for the visually impaired.

Home security systems are easily converted to incorporate options for occupants with hearing limitations. Smoke detectors can activate a large strobe light with virtually identical "wake and warn" efficiency to normal siren-type alarms, and doorbell chimes can incorporate a light.

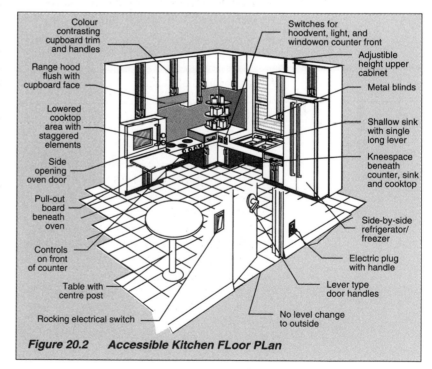

Figure 20.2   *Accessible Kitchen FLoor PLan*

20.2   *AFFORDABILITY*

One trend affecting new construction is the gradual decrease in the ability of Canadians to own their own home. Only half of all young families in urban centres own their own homes. Home ownership is still considered a good investment by all age groups and is preferred to renting.

The costs of construction and land have increased in most parts of the country. And housing costs as a percentage of income continue to increase. In the 1950s it took 25 percent of one family member's annual income to support a house. Today, it takes 40 to 50 percent of two incomes to support the same house.

Housing affordability can be improved through innovative subdivision design and alternative development standards, conversion of underutilized buildings, and innovative house designs and construction practices.

20.2.1   Smaller Lots and Higher Densities

One of the keys to affordability is changing the complex, multi-layered system of by-laws, regulations, and procedures that governs urban housing development. Collectively these regulations often delay development of, and even prevent new approaches to, innovative, lower cost housing projects.

And municipalities are seeing the need for change. Many Canadian municipalities are reevaluating their existing standards (see Figure 20.3) and identifying where changes could result in dollar savings without compromising safety, servicing or traffic flows. Changes being considered include:

- reducing road widths from 20 to 16 metres;
- allowing smaller lots, reduced frontage and setbacks;
- eliminating double trenching for services; and
- eliminating curbs and sidewalks.

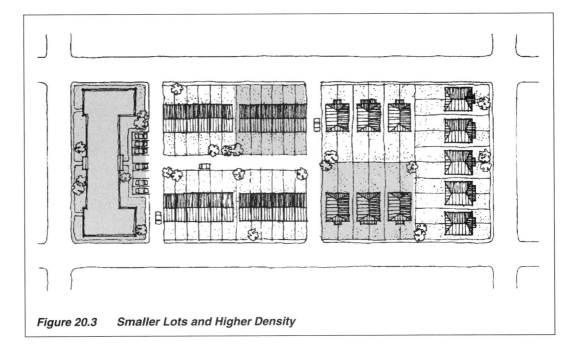

Figure 20.3   Smaller Lots and Higher Density

Based on 1,800 lots on an 85 acre parcel of land, one municipality discovered that by implementing alternative development standards, the cost of raw land could be reduced by $3,600 per unit, and servicing costs could be reduced by $3,600 per unit.

20.2.2 Grow and Charlie Homes

Most young families buy a starter home with the idea of selling it and upgrading to a better home. Some designs are working to change that concept.

With high-quality construction, reduced size and simplified construction procedures, construction costs can be reduced. Projects such as the Grow Home (see Figure 20.4a) incorporate a series of alternative design and construction concepts which can enhance affordability, including:

- total floor area reduced to 100m² (1,000 sq.ft.);
- bathroom placed immediately behind, or above, the kitchen for better efficiency of space and plumbing;
- a room that can be converted to a bedroom or sitting room, as required;
- the second floor left as open space;
- a master bedroom that can have a bathroom added later or can be separated into two bedrooms as the need arises;
- do-it-yourself components that may be added later such as cupboard doors to reduce initial costs and ongoing mortgage payments; and
- panelized components manufactured in the factory that can be assembled on site by two people in ten days.

Hamilton's Charlie house (see Figure 20.4b) was also designed to meet the needs of first-time home buyers, young professionals and young families. Its primary goal was to meet the needs of today's families but be flexible enough to expand from within should the need arise.

To tackle the debt load problems facing today's home buyer, the second floor of the 2,000-square-foot home can be converted to a rental unit to offset mortgage payments. When they require more space to raise families, owners can convert the second floor for their personal use. When the children leave home, the upper unit can be converted back to rental use or to quarters for grandparents or other family members while giving all parties privacy and independence.

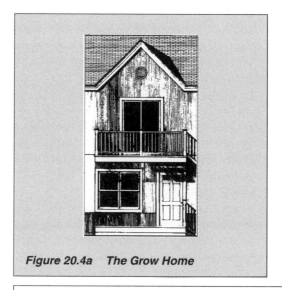

Figure 20.4a   The Grow Home

Figure 20.4b   The Charlie House

Alternatively, two households could pool resources: each would own half of the house and contribute to the downpayment and mortgage payments.

Key elements of design include:

- flexibility to allow for easy, inexpensive conversion, including future plumbing and electrical requirements, fire separations between floors, heating systems, and entry ways designed and built as part of the original structure;
- a floor plan that can be constructed as a row/town home or in three- or four-floor stacked configurations, with or without underground parking, depending on available land;
- carefully planned interior layout to maximize the use of available floor space — careful placement of doorways provides good traffic flow within the unit and cathedral ceilings on the second floor provide volume and create a feeling of more space;
- space saving techniques such as ceiling-mounted heating and air conditioning systems, stackable washer/dryers, closet organizers, and small yet efficient kitchens;
- space enhancing decor, such as bright colours, mirrors and window coverings, as well as attractively sized and placed furniture;
- balloon framing to allow movement of any interior wall; and
- R-2000 specs to keep the house's overall maintenance costs to a minimum.

20.2.3 Converting Underused Space

Canadian examples abound of successful conversions to living quarters of underused, vacant, or mothballed facilities. By converting schools, factories, mansions, and even old motels to affordable, accessible, and mixed-use living, some of the cultural and architectural heritage of this country is being preserved.

Schools, for example, are often well located within a city and community. The building fabric lends itself to the integration of contemporary dwelling-units within the given envelope. And because they were designed for public occupancy, they usually conform to the most demanding building and fire code requirements.

Even well-located office buildings are being converted into residential units.

20.3   HOME AUTOMATION

Home automation refers to the electronic linking of residential mechanical systems and devices including security, communications equipment, heating, ventilation, air conditioning, entertainment, lighting and appliances.

At the heart of a home automation system is a microprocessor-based controller which receives information from various mechanical systems and devices about their operational status (see Figure 20.5). The controller can use that information to alter the operation of some or all of the devices in accordance with a set of criteria preprogrammed by the homeowner. For example, a "good-bye" button at the front door allows the house to check itself as the occupants leave the home. Pressing the button triggers a sequence that arms the security system, checks to see that the phone answering machine is operational, that lights are out, windows shut, that the heating system is set back, and that appliances that shouldn't be left running are turned off.

This section provides a brief overview of home automation systems: how the technology can be used to meet the changing needs of home buyers, potential benefits that such systems can provide, where the technology stands now, and where it is headed.

20.3.1   Forces and Features

Housing is being shaped by a number of trends, including the changing workforce, an aging population, and concern for the environment. Home automation has the potential to be a simple, cost-effective way of meeting the changing needs of new home buyers.

Convenience

Two-income families have become a large part of the cultural landscape. Today, women comprise more than 40 percent of the work force. One result is that people now spend less time at home cooking, cleaning, and performing routine maintenance. Home automation systems have the potential to take the burden of many of these tasks off the shoulders of homeowners. Convenience is also an important factor for people who do spend a lot of time at home, such as seniors and the physically challenged. Both of these groups are growing in numbers at a time when there is a societal move away from institutionalization. For these people, home automation offers an easy-to-operate technical solution to many of the problems involved in living outside the health care system.

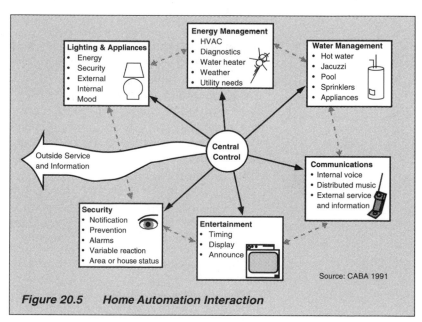

Figure 20.5   *Home Automation Interaction*

Safety and Security

For people who work away from home, and for people who spend most of their time at home, security has become a key residential issue. Home automation systems provide an advanced degree of safety and security — from turning a coffee pot off before it boils dry to informing an off-site security service if a window is forced open.

Energy Efficiency

Energy efficiency is still a major issue — for home builders, home buyers and society at large — particularly when the environmental ramifications of energy use are taken into account. Home automation systems have the ability to monitor and manage electrical use to provide optimum end-use efficiency. The systems may also help homeowners change their energy use patterns by providing feedback on when, where, and how they utilize electricity within the home.

For the builder, an investment in a home automation system may be a more cost-efficient means of achieving a desired level of energy efficiency than changes to the building envelope.

20.3.2   The State-of-the-Art

Home automation is still a relatively new innovation; however, a number of technologies have been developed, from single products to fully automated systems. At present, there is no single common communication system which allows devices and appliances to communicate with each other. There are several proprietary systems in existence, and more under development.

CABA identifies three categories within home automation: smart products, intelligent subsystems and home automation systems.

- Smart products are individual devices which have the ability to send/receive data or control messages from another product (e.g., a programmable VCR).

- An intelligent subsystem is an individual device, or group of devices, which uses information from other products to alter its own operation (such as an HVAC system that determines which zones require fresh air at any given moment).

- A home automation system is a central controller with the ability to control the functions of several devices and appliances.

At present, there are several home automation systems available to the Canadian builder. Each offers a different level of sophistication and obviously a different price.

20.3.3   The Future

For home automation systems to reach their potential, the most important development needed is a common communications protocol, i.e., the ability of one piece in one system to talk to a different piece in a different system. The large number of systems currently available tends to limit choices for inter-connecting products.

The near future should also see the development of faster and more intelligent equipment, allowing greater integration between devices that are on line.

Entertainment equipment will undoubtedly become the interface point between the user and the system. This interaction will be quick, easy and simple, most likely menu- or voice-driven.

Services provided by home automation systems will also reach outside the home, allowing such things as electronic postal service, newspaper information services, and billing and banking services through the television or telephone.

Prices should also drop substantially in the near future as capabilities and the number of products increase, and installation expertise becomes common.

20.4   WHAT'S NEXT?

Today's Canadian house bears little resemblance to that constructed as little as twenty years ago. Our houses are being constructed in significantly less time, and they are far superior in terms of energy efficiency, environmental responsibility and in providing high levels of occupant comfort, security and convenience.

The list of new technologies which have been adopted over the past twenty years includes HRV's, high efficiency heating and cooling systems, air tight construction practices, basement drainage membranes, high performance windows, materials with reduced toxicity and materials which are manufactured with recycled content.

The rate of change has been significant. And as the construction industry moves toward the turn of the century, this rate of technological change is likely to speed up.

New construction processes taking advantage of advances in computer technologies, building automation systems, and new products based on an evolving and improving understanding of building sciences will result in the provision of an improved product at increasingly affordable prices to Canadian consumers.

The Canadian Home Builders' Association, in conjunction with their federal partners at Natural Resources Canada, the Institute for Research in Construction of the National Research Council, and the Canada Mortgage and Housing Corporation, assumes a large role in keeping the residential construction industry abreast of the leading edge. Through initiatives such as the R-2000 Program and the Advanced Houses Program, CHBA works with its membership to demonstrate state-of-the-art housing technologies, and to communicate the results of these programs to the broader industry.

While the future of Canadian housing may unfold in a fashion which we can't currently predict, one can be certain that CHBA will continue to play a leadership role in the development and provision of the most up-to-date information to its members, and to other individuals involved in the residential construction industry — in Canada and around the world.

APPENDICES

METRIC CONVERSION

Imperial	Metric
2x2	38x38 mm
2x3	38x64 mm
2x4	38x89 mm
2x6	38x140 mm
2x8	38x190 mm
2x10	38x235 mm
2x12	38x286 mm
2'x8'	600 x 2400 mm
4'x8'	1200 x 2400 mm
16" OC	400 mm
24" OC	600 mm
1"	25.4 mm
1 1/2"	38 mm
2"	51 mm

Basic Metric Conversions	To get	multiply	by
length	mm	inches	25.4
	m	feet	0.3048
area	m^2	ft^2	0.0929
	yd^2		0.836127
volume	m^3	ft^3	0.028317
	yd^3		0.764555
flow	l/s	cfm	0.471947
energy	W	Btu/hr.	0.293072
insulation R value	RSI	R	0.1761
insulation R/unit thickness	RSI/mm	R/in.	0.00693
mass	kg	lb.	0.454
permeance (water vapour)	$ng/Pa.m^2s$	grain/ft^2hr (in.Hg)	57.5
resistance to moisture flow	$Pa.m^2s/ng$	ft^2h (in.Hg)/ grain	0.0174

Appendix *I*

INSULATION VALUES

The following pages include a chart showing the thermal resistance values of various building materials, and a discussion of how to calculate the thermal resistance of building assemblies.

Building Material	RSI/mm (R/in.) (average)Listed	RSI(R) for Thickness
Insulations		
Polyurethane boardstock	0.041 (6.0)	
Polyurethane (foamed in place)	0.041 (6.0)	
Extruded polystyrene boardstock	0.034 (5.0)	
Isocyanurate, sprayed	0.034 (5.0)	
High density glass fibre boardstock	0.029 (4.2) – 0.031 (4.5)	
Expanded polystyrene boardstock	0.026 (3.8) – 0.030 (4.4)	
Glass fibre roof board	0.028 (4.0)	
Cork	0.026 (3.8)	
Cellulose fibre, blown (settled thickness)	0.025 (3.6)	
Cellulose fibre, sprayed (settled thickness)	0.024 (3.5)	
Mineral fibre, batt	0.024 (3.5)	
Wood fibre	0.023 (3.3)	
Mineral fibre, loose fill	0.023 (3.3)	
Glass fibre, batt	0.022 (3.2)	
Glass fibre loose fill (poured in)	0.021 (3.0)	
Glass fibre loose fill (blown in)	0.020 (2.9)	
Fibreboard	0.019 (2.7)	
Mineral aggregate board	0.018 (2.6)	
Wood shavings	0.017 (2.5)	
Expanded mica (vermiculite, zonolite, etc.)	0.016 (2.3)	
Compressed straw board	0.014 (2.0)	

Building Material	RSI/mm (R/in.) (average)	RSI(R) for Thickness Listed
Structural Materials		
Cedar logs and lumber	0.0092 (1.3)	
Softwood lumber (except cedar)	0.0087 (1.3)	
Concrete		
480 kg/m³ - 2400 kg/m³ (30-150 lb./ft.³)	0.0069 - 0.00045 (0.06-1.0)	
Concrete block (3 oval core)		
sand and gravel aggregate		
100-300 mm (4-12 in.)		0.12-0.22 (0.68-1.25)
Cinder aggregate		
100 - 300 mm (4-12 in.)		0.20-0.33 (1.14- 1.9)
Lightweight aggregate		
100-300 mm (4-12 in.)		0.26-0.40 (1.5-2.3)
Common brick		
Clay or shale, 100 mm (4 in.)		0.07 (0.4)
Concrete mix, 100 mm (4 in.)		0.05 (0.3)
Stone (lime or sand)	0.00060 (0.087)	
Steel	0.000022 (0.003)	
Aluminum	0.0000049 (0.0007)	
Glass (no air films), 3-6 mm (1/8 -1/4 in.)		0.01 (0.06)
Air		
Enclosed air space (nonreflective)		
Heat flow down, 20-100 mm (1-4 in.)		0.18 (1.0)
Heat flow horizontal, 20-100 mm (1-4 in.)		0.17 (1.0)
Heat flow up, 20-100 mm (1-4 in.)		0.15 (0.9)
Air surface films		
Outside air film (moving air)		0.03 (0.2)
Inside air film (still air)		
Horizontal, heat flow down		0.16 (0.9)
Vertical, heat flow horizontal		0.12 (0.7)
Horizontal, heat flow up		0.11 (0.6)
Sloping 45°, heat flow up		0.11 (0.6)
Attic air film		0.08 (0.5)
Roofing		
Asphalt roll roofing		0.03 (0.2)
Asphalt shingles		0.08 (0.5)
Wood shingles (cedar shakes)		0.17 (1.0)
Built-up membrane (hot mopped)		0.06 (0.3)
Sheathing Materials		
Softwood plywood	0.0087 (1.3)	
Mat-formed particleboard	0.0087 (1.3)	
Insulating fibreboard sheathing	0.017 (2.5)	
Gypsum sheathing	0.0062 (0.9)	
Sheathing paper	0.0004 (0.06)	
Asphalt-coated kraft paper vapour diffusion retarder (VDR)	negligible	
Polyethylene VDR	negligible	

Building Material	RSI/mm (R/in.) (average)	RSI(R) for Thickness Listed
Cladding Materials		
Fibreboard siding		
Medium-density hardboard,		
9.5 mm (3/8 in.)		0.10 (0.6)
High-density hardboard,		
9.5 mm (3/8 in.)		0.08 (0.5)
Softwood siding (lapped)		
Drop, 18x184 mm (1x8 in.)		0.14 (0.8)
Bevel, 12x184 mm (1/2x8 in.)		0.14 (0.8)
Bevel, 19x235 mm (1x10 in.)		0.18 (1.0)
Plywood, 9 mm (3/8 in.)		0.10 (0.6)
Wood shingles		0.17 (1.0)
Brick (clay or shale), 100 mm (4 in.)		0.08 (0.5)
Brick (concrete and sand (lime)),		
100 mm (4 in.)		0.05 (0.3)
Stucco, 25 mm (1 in.)	0.0014 (0.20)	0.03 (0.2)
Metal siding		
Horizontal clapboard profile		0.12 (0.7)
Horizontal clapboard profile		
with backing		0.25 (1.4)
Vertical V-groove profile		0.12 (0.7)
Vertical board and batten profile		negligible
Interior Finish		
Gypsum board, gypsum lath,		
12.7 mm (1/2 in.)	0.0062 (0.89)	0.08 (0.5)
Gypsum plaster		
Sand aggregate, 12.7 mm (1/2 in.)	0.0014 (0.20)	0.02 (0.1)
Lightweight aggregate,		
12.7 mm (1/2 in.)	0.0044 (0.63)	0.06 (0.3)
Plywood, 7.5 mm (5/16 in.)	0.0093 (1.34)	0.07 (0.4)
Hardboard (standard), 6 mm (1/4 in.)	0.0053 (0.76)	0.03 (0.2)
Insulating fibreboard, 25 mm (1 in.)	0.017 (2.45)	0.42 (2.4)
Drywall, gypsum board, 12.7 mm (1/2 in.)	0.0061 (0.88)	0.08 (0.5)
Flooring		
Maple or Oak (hardwood), 19 mm (3/4 in.)	0.0063 (0.91)	0.12 (0.7)
Pine or Fir (softwood), 19 mm (3/4 in.)	0.0089 (1.28)	0.17 (1.0)
Plywood, 16 mm (5/8 in.)	0.0088 (1.27)	0.14 (0.8)
Mat-formed particleboard, 16 mm (5/8 in.)	0.0088 (1.27)	0.14 (0.8)
Wood fibre tiles, 12.7 mm (1/2 in.)	0.016 (2.31)	0.21 (1.2)
Linoleum or tile (resilient), 3 mm (1/8 in.)	0.01 (0.06)	
Terrazzo, 25 mm (1 in.)	0.00056 (0.08)	0.01 (0.06)
Carpet, typical thickness		
Fibrous underlay		0.37 (2.1)
Rubber underlay		0.23 (1.3)

Calculating the Thermal Resistance of Building Assemblies

The wood framing typically accounts for about 15 per cent of the wall surface area in an assembly with 38x140 mm (2x6 in.) studs at 600 mm (24 in.) o.c. Therefore, the thermal resistance of the wall assembly can be calculated as follows (see Figure I.1).

Construction	Resistance Between Framing	(RSI(R)) At Framing
1. Outside surface (24 kmh/15 mph wind)	0.03 (0.17)	0.03 (0.17)
2. Siding, wood, 12.7 mm x 200 mm (1/2 in. x 8 in.) lapped (average)	0.14 (0.81)	0.14 (0.81)
3. Sheathing, 12.7 mm (1/2 in.) mat-formed particleboard	0.11 (0.65)	0.11 (0.65)
4. RSI 3.5 (R-20) batt-type insulation	3.5 (20)	—
5. Nominal 38x140 mm (2x6 in.) wood stud	—	1.3 (7.5)
6. Gypsum wallboard, 12.7 mm (1/2 in.)	0.08 (0.45)	0.08 (0.45)
7. Inside surface (still air)	0.12 (0.68)	0.12 (0.68)
Total Thermal Resistance - RSI(R)	3.98 (22.76)	1.78 (10.26)

$$\frac{1}{RSI\ (R)_{av.}} = (\text{fraction of wall surface between framing}) \times \frac{1}{(\text{insulating value between framing})}$$

$$+ (\text{fraction of wall surface at framing}) \times \frac{1}{(\text{insulating value at framing})}$$

$$\frac{1}{RSI_{av.}} = (0.85) \times \frac{1}{(3.98)} + (0.15) \times \frac{1}{(1.78)} = 0.2978 \qquad RSI_{av.} = \frac{1}{0.2987} = 3.36$$

$$\frac{1}{R_{av.}} = (0.85) \times \frac{1}{(22.76)} + (0.15) \times \frac{1}{(10.26)} = 0.052 \qquad R_{av.} = \frac{1}{0.052} = 19.23$$

The insulating value of ceiling and floor assemblies can be calculated in a similar manner. The wood framing typically accounts for about 10 per cent of the surface of these assemblies.

Figure 1.1 *Calculating the Thermal Resistance of Building Assemblies*

Effective Thermal Characteristics of Common Building Assemblies

Abbreviations and symbols

Effective RSI   – effective thermal resistance of the overall assembly, including insulation, sheathing and finishing materials, air films and thermal bridging of framing members;

Effective Added RSI   – thermal resistance of an insulated assembly added to the basement wall or floor, including insulation, finishing materials and thermal bridging of framing members (excluding air films, earth and concrete or plywood sheathing of treated wood foundations);

U–value   – overall heat transmittance (for windows, this includes centre glass, edge of glass and framing, as well as air films);

SHGC   – solar heat gain coefficient;

ER   – energy rating of the window assembly;

EPS II   – expanded polystyrene (bead board) type 2;

XTPS II   – extruded polystyrene type 2;

XTPS IV   – extruded polystyrene type 4;

Attics		Effective RSI
Description		
Truss @ 600 mm oc;	RSI 3.5 blown insulation	3.7
	RSI 4.9 blown insulation	5.1
	RSI 5.6 blown insulation	5.8
	RSI 7.0 blown insulation	7.2
	RSI 8.8 blown insulation	9.1
	RSI 10.6 blown insulation	10.9

Roof-joist Type Roofs	Effective RSI
Description	
184 mm Joist @ 400 mm oc; RSI 3.5 batt	3.25
235 mm Joist @ 600 mm oc; RSI 3.5 batt	3.38
286 mm Joist @ 600 mm oc; RSI 3.5 batt	3.54
184 mm Joist @ 400 mm oc; RSI 3.5 batt; + 0.7 EPS II	4.02
235 mm Joist @ 600 mm oc; RSI 4.9 batt	4.61
286 mm Joist @ 600 mm oc; RSI 4.9 batt	4.77
184 mm Joist @ 400 mm oc; RSI 3.5 batt; + 1.4 EPS II	4.77
286 mm Joist @ 600 mm oc; RSI 5.4 batt	5.21
235 mm Joist @ 600 mm oc; RSI 4.9 batt; + 1.4 EPS II	6.29
286 mm Joist @ 600 mm oc; RSI 7 batt	6.55
406 mm I–Joist @ 600 mm oc; RSI 7 batt	7.11
286 mm Joist @ 600 mm oc; RSI 7 batt; + 1.4 EPS II	8.12
406 mm I–Joist @ 600 mm oc; RSI &7.8 cellulose	8.13
406 mm I–Joist @ 600 mm oc; RSI 8.4 batt	8.50

Wood Frame Walls	Effective RSI
Description	
38x89 @400mm oc; RSI 2.1 + RSI 1.32 XTPS II	3.43
38x89 @400mm oc; RSI 2.1 + 1.6 polyisocyanurate	3.72
38x140 @400mm oc; RSI 3.4 batt (3.52 compressed)	3.06
38x140 @400mm oc; RSI 3.4 + .88 XTPS II	3.97
38x140 @400mm oc; RSI 3.4 + 0.77 semi-rigid	3.85

Wood Frame Walls continued	Effective RSI
38x140 @400mm oc; RSI 3.4 + .79 polyisocyanurate	3.87
38x140 @400mm oc; RSI 3.4 + 1.05 EPS II	4.16
38x140 @400mm oc; RSI 3.4 + 1.18 semi-rigid	4.29
38x140 @400mm oc; RSI 3.4 + RSI 1.32 XTPS II	4.44
38x140 @400mm oc; RSI 3.4 + 1.6 polyisocyanurate	4.73
38x89 @400 & 38x64 @ 600; RSI 2.1 + 4.9 batts	6.79

Basement Walls	Effective Added RSI
Description:(Building paper assumed for framed assemblies; subtract 0.01 if not used.)	
Full depth: 38 mm EPS II @ full	1.14
Full depth: 51 mm EPS II @ full	1.50
Full depth: 64 mm EPS II @ full	1.86
Full depth: 76 mm EPS II @ full	2.20
Full depth: 89 mm EPS II @ full	2.55
Full depth: 100 mm EPS II @ full	2.91
Full depth: 89 mm XTPS IV @ full	3.18
Full depth: 100 mm XTPS IV @ full	3.63
Full depth: 38 x 89 @ 600 mm oc; RSI 2.1 batt	1.90
Full depth: 38 x 89 @ 600 mm oc; RSI 3.5 batt	3.39
Full depth: RSI 2.1 batt horizontal + framed RSI 2.1 vertical	4.10
Full depth: 38 x 89 @ 600 mm oc; RSI 4.9 batt	4.82
Full depth: 38 x 89 @ 600 mm oc; RSI 5.4 batt	5.32
Full depth: 38 x 89 @ 600 mm oc; RSI 7.0 batt	6.93

Floors in Contact with the Ground	Effective Added RSI
Description	
25 mm XTPS IV	0.88
38 mm XTPS IV	1.32
51 mm XTPS IV	1.76
76 mm XTPS IV	2.63

Operable Windows	U-value	SHGC	ER
Description Format: **number of glazings**, coating or film (if any), fill (if not air), frame, spacers			
double, vinyl frame	2.54	0.46	-23.2
double, low-e, vinyl frame	2.01	0.45	-12.5
triple, clear, wood frame	2.01	0.46	-11.6
double, low-e, argon, vinyl frame	1.88	0.45	-9.7
double, low-e, argon, vinyl frame, insulated spacer	1.77	0.47	-5.4
double, low-e, argon, foam-filled vinyl frame insulated spacer	1.61	0.45	-2.5
triple, low-e(1), argon(1), foam-filled vinyl frame, insulated spacer	1.25	0.41	1.9
triple, low-e(2), argon(2), foam-filled vinyl frame, insulated spacer	1.06	0.37	3.0
triple, low-e(2), argon(2), fibre glass frame, insulated spacer	1.06	0.40	5.7

Fixed Windows (no sash)	U-value	SHGC	ER
Description			
double, vinyl frame	2.75	0.64	-13.7
double, low-e, vinyl frame	2.14	0.62	-2.5
triple, clear, wood frame	1.99	0.62	0.9
double, low-e, argon, vinyl frame	1.92	0.59	0.2
triple, low-e, argon, vinyl frame	1.41	0.56	9.4
triple, low-e(1), argon(1), foam-filled vinyl frame, insulated spacer	1.30	0.56	11.8
triple, low-e(2), argon(2), foam-filled vinyl frame, insulated spacer	1.05	0.50	13.2
triple, low-e(2), argon(2), fibre glass frame, insulated spacer	0.93	0.49	14.8

Appendix *II*

AIR and VAPOUR PERMEANCE

Vapour Permeance

Materials considered sufficiently resistant to the flow of water vapour to be used as vapour diffusion retarders (VDRs) are rated in terms of their *permeance*. The lower the perm rating of a material, the more effectively it will retard diffusion.

In SI (metric) units, a perm represents a transfer of one nanogram of water per square metre of material per second under a pressure difference of one pascal. In imperial units, this corresponds to one grain (0.002285 oz.) of water per square foot of material per hour under a pressure difference of one inch of mercury (1.134 ft. of water).

There are two classes of VDRs. A *Type I* VDR has a permeance of 14.375 metric perms (0.25 perms in imperial units) or less. A *Type II* VDR has a permeance of 43.125 (0.75) perms or less before aging and 57.5 (1.0) perms or less after aging. Any material with a perm rating higher than 57.5 (1.0) perms is *not* a VDR.

Air permeance

Air leakage is considered to be of great importance when it comes to the performance of a building. Air leakage is responsible for major transfers of heat, water vapour, and air born pollutants. Many materials have the characteristics required of air barriers. The air permeance of a building material is defined as the rate of air flow (L/s), per unit area (m2) and, per unit static pressure differential (Pa). The resistance to air flow provided by a building material (R) is the reciprocal of the air permeance.

Material	VAPOUR PERMEANCE Metric (ng/Pa.m²s)	Imperial (grain/ft²hr (in.Hg))
Common Vapour Diffusion Retarders		
0.03 mm (1 mil) aluminum foil	0	0.00
0.01 mm (0.35 mil) aluminum foil	2.9	0.05
0.15 mm (6 mil) polyethylene	3.4	0.06
0.10 mm (4 mil) polyethylene	4.6	0.08
0.05 mm (2 mil) polyethylene	9.1	0.16
asphalt kraft paper facing	17	0.30
0.05 mm (2 mil) unplasticized polyvinylchloride	39	0.68
0.03 mm (1 mil) polyester	42	0.73
Paint and Wallpaper		
2 coats asphalt paint on plywood	23	0.40
2 coats aluminum paint	17-29	0.30-0.50
1 coat latex VDR paint	26	0.45
3 coats white lead/oil paint on wood siding	17-57	0.30-0.99
3 coats white lead/zinc oxide/oil paint on wood	51	0.88
2 coats oil-based paint on plaster	91-172	1.58-2.99
Insulations		
25 mm (1 in.) extruded polystyrene	23-92	0.40-1.60
25 mm (1 in.) polyurethane	69	1.20
25 mm (1 in.) expanded polystyrene	115-333	2.00-5.79
100 mm (4 in.) rock wool	1666	28.97
100 mm (4 in.) cellulose fibre	1666	28.97
100 mm (4 in.) glass fibre wool	1666	28.97
Other Building Materials		
100 mm (4 in.) glazed tile masonry	6-9	0.10-0.16
3 mm (1/8 in.) asbestos-cement board with oil paint	17-29	0.30-0.50
19 mm (3/4 in.) board (wood)	17-232	0.30-4.03
6.4 mm (1/4 in.) CDX plywood	40	0.70
100 mm (4 in.) brick	46	0.80
200 mm (8 in.) concrete block	137	2.38
6.8 kg (15 lb.) tar paper	230	4.00
3 mm (1/8 in.) hardboard (standard)	630	10.96
19 mm (3/4 in.) plaster on metal lath	860	14.96
spunbonded polypropylene	884	15.37
12.7 mm (1/2 in.) insulating board	1150-2875	20.00-50.00
builder's sheathing paper	1170	20.35
gypsum drywall	2860	49.74
spunbonded polyolefin	3646	63.41

AIR PERMEANCE	
Material	**Measured Leakage at 75 Pa.** L/s-m2
Materials Showing a Non-measurable Air Flow	
0.15 mm (6 mil) polyethylene	0.00
1.50 mm (1/16 in.) smooth surface roofing membrane	0.00
2.70 mm (3/32 in.) modified bituminous torch on grade membrane (glass fibre mat)	0.00
0.03 mm (1 mil) aluminum foil	0.00
1.50 mm (1/16 in.) modified bituminous self-adhesive membrane	0.00
2.70 mm (3/32 in.) modified bituminous torch on grade membrane (polyester reinforced mat)	0.00
9.50 mm (7/16 in.) plywood sheathing	0.00
38.0 mm (1.5 in.) extruded polystyrene	0.00
25.4 mm (1 in.) foil back urethane insulation	0.00
25.4 mm (1 in.) phenolic insulation board	0.00
50.8 mm (2 in.) phenolic insulation board	0.00
12.7 mm (1/2 in.) cement board	0.00
12.7 mm (1/2 in.) foil back gypsum board	0.00
Materials Having a Measurable Air Flow	
8.00 mm (3/8 in.) plywood sheathing	0.0067
16.0 mm (5/8 in.) flakewood board	0.0069
12.7 mm (1/2 in.) gypsum board (M/R)	0.0091
11.0 mm (7/16 in.) flakewood board	0.0108
12.7 mm (1/2 in.) particle board	0.0155
reinforced non-perforated polyolefin	0.0195
12.7 mm (1/2 in.) gypsum board	0.0196
15.9 mm (5/8 in.) particle board	0.0260
3.00 mm (1/8 in.) hardboard (standard)	0.0274
25.0 mm (1 in.) expanded polystyrene - type II	0.1187
spunbonded polyolefin film	0.1776
13.6 kg (30 lb.) roofing felt	0.1873
6.8 kg (15 lb.) non-perforated asphalt felt	0.2706
6.8 kg (15 lb.) perforated asphalt felt	0.3962
glass fibre rigid insulation board with spunbonded polyolefin film on one face	0.4880
11.0 mm (7/16 in.) plain fibreboard	0.8223
11.0 mm (7/16 in.) asphalt impregnated fibreboard	0.8285
spunbonded polypropylene film	3.2186
0.10 mm (4.00 mil) type II perforated polyethylene	3.2307
0.10 mm (4.00 mil) type I perforated polyethylene	4.0320
25 mm (1 in.) expanded polystyrene - type I	12.2372
tongue and groove planks	19.1165
152 mm (6 in.) glass fibre wool insulation	36.7327
vermiculite insulation	70.4926
cellulose insulation (spray on)	86.9457

Appendix *III*

VENTILATION SYSTEM INSTALLATION CHECKLIST

Although this checklist has been specifically developed for an HRV installation, it can be applied generally to any type of ventilating system.

General

The installer should:

- Be trained and registered by the Heating, Refrigeration and Air Conditioning Institute of Canada (HRAI).

- Perform the installation in accordance with the CAN/CSA-F326-M91, *Residential Mechanical Ventilation Systems* (if an HRV is being installed).

- Provide you or the homeowner with all system documentation and maintenance and operation manuals.

Ventilating equipment should be located where it will be:

- In the heated interior of the house.

- Away from noise-sensitive living areas (dining room, living room, bedroom, etc.).

- Close to an outside wall to minimize insulated duct runs.

- Convenient to a drain (for condensate) and to an electrical supply.

- Easily accessible for maintenance purposes.

- Away from hot chimneys, electrical panels, and other possible fire sources.

Ventilator Installation

The ventilator should be installed:

– With "vibration isolators" (such as rubber feet and a short section of flexible duct between the unit and the main ductwork).

– With a control system capable of providing continuous low-speed operation plus humidity-controlled and manually controlled high speed override.

– With air flow measuring stations installed in the ducts such that all the supply and exhaust air can be measured; the stations should be installed on the warm side of any HRV.

"Cold Side" Ducts

Two "cold side" ducts should connect the HRV to the outside, with one bringing in fresh air and the other exhausting stale air. The ducts should:

– Be as short and straight as possible.

– Contain no kinks or depressions where condensate water might accumulate.

– Be insulated with a minimum of RSI 0.4 (R-2) insulation.

– Be sealed from end to end (outside the insulation) with a vapour diffusion retarder (VDR).

– Have their VDRs sealed to the VDR of the house envelope.

– Be clearly marked as to which is for fresh air and which for exhaust air.

– Terminate in two accessible rain hoods, each equipped with a 6 mm (1/4 in.) or coarser wire mesh bird screen. If finer insect screens are used, they must be accessible for regular cleaning.

The cold side fresh air inlet should:

– Be located at least 450 mm (18 in.) above ground (more if exposed to blockage by snow, etc.) and at least 1800 mm (72 in.) away from the exhaust outlet.

– Be located away from sources of contamination such as carports, driveways, garages, bushes, tall grass, garbage, dryer vents, central vacuum exhausts, other vents, oil fill pipes, gas regulators, etc.

– Be equipped with a filter in the duct or at the inlet, if the HRV does not have a filter.

The cold side exhaust outlet should:

– Be located away from walkways and other areas where ice accumulation could be a problem (since moisture in the exhaust air may condense and freeze).

– Be terminated outdoors (*not* in a garage or attic), at least 200 mm (8 in.) above ground (more if practical).

"Warm Side" Ducts: General

"Warm side" ducts should connect the HRV to the house, with one duct system distributing fresh air while a second collects exhaust air.

– Either an exhaust or a fresh air duct should be provided to every room.

– Duct systems should be designed to minimize their length and complexity (turns, right angles, etc.).

– Rigid ducts should be used wherever possible in preference to flexible ducts, since the ribbing in flexible ducts doubles the resistance to air flow.

– Duct runs in unheated spaces and in exterior walls should be avoided.

– Duct joints should be securely fastened and taped.

"Warm Side" Exhaust Ducts

Exhaust ducts should:

– Run from such areas as bathrooms and kitchens, and other areas where contaminants are generated.

– Contain a filter in the duct or at the grilles, if the HRV does not have a filter.

In the kitchen:

– The range hood must not be connected to the HRV.

– The general kitchen exhaust (connected to the HRV) must be at least 1200 mm (48 in.) horizontally removed from the cooking surface.

– Unless a charcoal filter grille is used, the kitchen exhaust must be ducted in sheet metal.

In the laundry room:

– Clothesdryers should be vented directly to the outside.

"Warm Side" Fresh Air Ducts

Fresh air can be distributed throughout the house by a separate system of ducts or by a forced air furnace.

If a separate system of ducts is used:

– Registers should be positioned to minimize discomfort from cold drafts (e.g., at ceiling level, in hallways, etc.).

If a forced-air furnace is used to distribute fresh air:

– An HRV integrated with a two-speed fan heat distribution system should dump fresh air 100 to 300 mm (4 to 12 in.) from an opening in the furnace cold air return, and the opening in the return should be at least 2000 mm (80 in.) from the plenum if an indirect connection is being used.

– A direct connect can only be used if the heat distribution system uses a single-speed circulating fan and the heating system and HRV are designed for the application.

Verification of Operation

Once the unit is operational, the installer should:

– Where the HRV has its own distribution ductwork, check and adjust flows in all ducts to ensure that sufficient air flow is provided to all rooms when the doors are closed.

– Adjust the total air flow to provide the continuous ventilation rate as specified.

– Balance the flow, where appropriate, to equalize the flow rates of the fresh air and the exhaust air.

– Ensure that the controls work as designed. This includes the (de)humidistat and an interval timer and/or manual override.

– Mark the minimum ventilation settings on all dampers and speed controls so that they may be readjusted after the installation if necessary.

Appendix **IV**

R-2000 HOME PROGRAM TECHNICAL REQUIREMENTS

1. PURPOSE

1.1 The R-2000 Home Program Technical Requirements provide the basis for the design and construction of new residential buildings which will be more efficient in the use of energy, improve the level of indoor air quality and address the environmental responsibility aspects of house construction and operation. This can be done through the use of such features as higher envelope insulation, airtightness, orientation, heating, cooling and ventilation as well as the selection of materials and components for use in the home.

1.2 The Requirements are intended to allow flexibility in the selection of thermal resistance values, lighting equipment, appliances (when they are supplied by the builder), cooling, ventilation and heat recovery equipment, and heating equipment for both space and domestic hot water to meet a total home annual energy consumption target for a particular location.

1.3 The Ventilation Systems and Equipment along with the Indoor Air Quality sections of these Technical Requirements are intended to ensure acceptable levels of air quality, adequate venting of combustion products, and control of indoor humidity levels for safety and health.

1.4 The Requirements provide a method by which achievement of the annual energy target can be assessed at the design stage through the evaluation of plans and specifications.

1.5 The Environmental Requirements and Indoor Air Quality Requirements are intended to allow flexibility in the selection of products and components used in the building to promote improved levels of indoor air quality and so the building has a reduced impact on the environment during the construction process and while in operation.

2.   SCOPE

2.1   These Requirements apply to ground related low rise housing as per Part 9 of the *National Building Code of Canada* ("N.B.C.") (except as noted in 2.5) to be constructed under the R-2000 Home Program by Registered Builders.

2.2   All houses constructed to these Requirements must comply with local and provincial code(s) or in the absence of such code(s) legislation, to the requirements of the current edition of the *National Building Code of Canada*.

2.3   These Requirements are <u>in addition to</u> the requirements of local, provincial or National Building Codes and Standards.

2.4   Recognizing that the R-2000 Home Program is designed to encourage innovation and creativity, these R-2000 Technical Requirements may be amended if changes are identified in advance and agreed to by Canadian Home Builders' Association ("CHBA") and Energy, Mines and Resources Canada ("EMR").

2.5   The minimum standard of performance for a multiple unit building with common heated areas, ventilation system or heating system will be agreed upon (between CHBA, EMR, and the Builder) prior to construction.

2.6   CHBA has the responsibility for ensuring equivalency and the authority to accept equivalent methods.

3.   MINIMUM ENVELOPE REQUIREMENTS

3.1   The following minimum insulation levels shall apply:

Degree Day Zone	Exterior Above Grade Walls		Exterior Below Grade Walls		Insulated Ceiling and Attics	
	RSI	R	RSI	R	RSI	R
Up to 3500	2.8	16	1.8	10	4.7	27
3501 – 6000	3.6	20	1.8	10	5.6	32
6001 – 8000	4.2	24	2.8	16	6.4	36
8001 and over	4.7	27	3.6	20	7.1	40

3.2   Windows shall be at least double glazed with a minimum air space thickness of 12.5 mm (0.5 in.) between panes. Metal window frames must be thermally broken except in warm areas where the January 2 $^{1/2}$ % Design Temperature is not less than -12 degrees Celsius.

3.3   Either the Normalized Leakage Area (NLA) of the building envelope shall be no greater than 0.7 cm²/m² (1 sq. in./100 ft²) area of the building envelope or the air change rate at 50 Pa shall be no greater than 1.5 ACH. Either shall be determined according to CAN/CGSB2-149.10-M86 "Determination of the Airtightness of Building Envelopes by the Fan Depressurization Method" or by equivalent methods as approved by CHBA. Additional stipulations on test procedure in 3.4 and 3.5 will supersede these Requirements.

3.4   A dwelling unit shall be tested individually without fan depressurization of any adjacent heated space. Envelope area will include that of building components separating a dwelling unit from other dwelling units, heated space and/or the outdoors.

3.5   Wood stoves and fireplace flues may be sealed for the airtightness test. Temporary sealing of sumps during airtightness tests is not allowed.

4. *VENTILATION SYSTEMS AND EQUIPMENT*

4.1   Mechanical ventilation systems shall be designed by an HRAI registered Ventilation Installer or Ventilation Designer. Mechanical ventilation systems shall be installed by an HRAI registered installer in accordance with CAN/CSA-F326-M91, Residential Mechanical Ventilation Systems.

4.2   Heat Recovery Ventilators (HRV) permitted for use shall be certified by the Home Ventilating Institute (HVI).

5. *COMBUSTION EQUIPMENT*

5.1   All gas, propane, and oil-fired space and water heating equipment shall have direct vent (sealed) or induced or forced draft venting systems. Induced draft or forced draft vented systems shall be capable of positive shut-down in the case of venting failure.

> **5.1.1**   The following mechanically vented induced draft (power vented) gas and propane appliances will not be subject to a pressure decrease limit in accordance with CSA F326:
>
> – furnaces without standing pilot lights
> – domestic hot water heaters without standing pilot lights
> – clothes dryers without standing pilot lights
> – fireplaces with non-operable doors and without standing pilot lights

5.2   All wood burning appliances including, fireplaces, woodstoves and pellet stoves must be certified as meeting the requirements of either:

> 1)   CSA B415.1-M92 Performance Testing of Stoves, Inserts and Low to Medium Burn Rate Factory Built Fireplaces, or
> 2)   The United States Environmental Protection Agency (EPA) wood burning appliance standards (1990), CFR Part 60.

5.3   Gas and propane fireplaces must be either:

> 1)   Direct-vent (sealed); either top-, or rear-vented
> 2)   Power-venting.

5.4   Fireplaces and chimneys on exterior walls shall be constructed to maintain continuity of the house air barrier and shall be insulated with non-combustible insulation. Air cooled chimneys are not permitted.

5.5   Where ducts that supply combustion air to combustion appliances pass through conditioned space they must be insulated with a minimum of RSI 0.4 (R-2) and have a vapour barrier to avoid condensation on the duct.

5.6   No unvented combustion appliance shall be installed unless specific provision is made to exhaust the products of combustion to the outside.

5.7   Electric domestic water heaters shall either have factory-installed insulation with a minimum RSI of 1.8 (R-10) or have a standby loss of 80 watts or less for a 270 L (60 gal.) tank measured in accordance with CSA C191 Series M1983.

5.8   Gas-fired domestic water heaters shall have a standby loss of 3.5% or less measured in accordance with CGA CAN-4.1-77 Gas Fired Automatic Storage Type Water Heaters with Inputs Less than 75000 BTU.

6.   ENERGY PERFORMANCE TARGETS

6.1   The annual energy consumption target for an R-2000 Home is determined from the following equation:

Annual Energy Target: $Q_s + Q_w$
where:

Q_s = space heating energy consumption target

Q_w = domestic hot water energy consumption target

6.1.1   The annual space heating energy consumption target is calculated using the equation:

$Q_s = S*F_u*F_v* (5+55*DD/6000) *V/2.5$
where:

S = 4.5 megajoules (MJ) for fuel-fired space heating systems

S = 1.0 kilowatt hours (kWh) or 3.6 megajoules (MJ) for electric space heating systems

F_u = .8 (a utilities consumption factor that compensates for the reduced space heating requirement due to increasing the default utilities consumption from 16 to 24 kWh/d).

F_v = Volume factor (provides a credit for houses with a heated volume smaller than 590 m^3 and a modifier for houses larger than 590 m^3).

F_v = $1.65 - 0.00145*V + 0.00000059*V^2$ for V up to 1200 m^3

F_v = 0.76 for V greater than 1200 m^3

DD = Celsius heating degree days for the locality

V = Interior heated volume, including basement, in cubic metres

6.1.2   The annual domestic hot water heating energy consumption target is calculated using the equation:

$Q_w = 4745*W$
where:

W = 2.0 kilowatt hours (kWh) or 7.2 megajoules (MJ) for fuel-fired DHW systems

W = 1.075 kilowatt hours (kWh) or 3.87 megajoules (MJ) for electric DHW systems

6.1.3   Reserved for future inclusion of the annual space cooling energy target.

6.1.4   Reserved for future inclusion of the annual lighting energy target.

6.1.5   Reserved for future inclusion of the annual appliance energy target.

6.2   The current authorized version of the HOT-2000 computer analysis program is used as the basis for determining compliance with the annual energy target. All HOT-2000 computer analysis are to be done in accordance with "HOT-2000 Design Approval Procedures and Guidelines".

7. LIGHTS AND APPLIANCES

7.1 Builders including appliances with the sale of a home shall offer the home owner the option of selection of appliances in the upper 33% of the Energuide rating for that appliance category.

7.2 An energy credit will be allowed for builder installed, high efficiency lamps (40 lumens/watt minimum). The credit will be given for each lamp, up to a total of 4, installed in any combination of the following rooms: bathrooms, bedrooms, hallways, kitchen, utility room, or other finished rooms.

7.3 An energy credit will be allowed for electronically commutated motors (ECM) used for the furnace fan.

(Note: Further study on clauses 7.2 and 7.3 is being undertaken to determine the credit size and methodology to implement the credit.)

8. INDOOR AIR QUALITY

8.1 At least two of the following options, which refer to materials used only inside the air barrier or air/vapour barrier, shall be selected:

    (i)   Carpeting; carpeting (except as noted) shall cover no more than 50% of the interior floor area. The interior floor area includes the basement floor area. The following floor coverings are exempt; wool or cotton area rugs and carpeting that has latex-free backing. These floor coverings shall not be glued to the floor and cannot have underpad.

    (ii)   Air Filtration; a medium efficiency air filter with a minimum 10% ASHRAE average dust spot efficiency shall be installed where air circulating heating, cooling and ventilation systems are used.

    (iii)   Paint and Varnishes; all liquid coatings used indoors except on wood floors, are to be water based, interior type or meet or exceed Environment Canada Environmental Choice standards. Prefinished items are allowed.

    (iv)   Flooring Adhesives; all finish flooring adhesives shall be either water dispersion, low toxicity formulations or pre-adhesive types.

    (v)   Kitchen Cabinets and Bathroom Vanities; cabinets and vanities shall be solid wood or if made from manufactured wood products shall be made from formaldehyde-free fibre board, or particle board meeting the E-1 European Standard or the HUD Standard, 24 CFR Part 3280.308, or have all exposed surfaces sealed with an Environmental Choice water based sealer or a low toxicity sealer.

    (vi)   Wood Flooring; all liquid coatings used on wood flooring shall meet or exceed Environment Canada Environmental Choice standards or be prefinished.

9. ENVIRONMENTAL FEATURES/ECO-MANAGEMENT

9.1   Water Conservation
Where the following fixtures are installed, they shall meet the following criteria:

    toilets: water saver or ultra-low flush units; 13.25 litres/flush or less

    showers: low flow shower heads; less than 9.8 litres/min. (2.15 imp. GPM) when tested at 5.5 kg/cm^3 (80 psi)

    faucets: lavatory and kitchen faucets less than 8.3 litres/min. (1.84 imp. GPM) when tested at 4.1

9.2   Materials Conservation

The dwelling units shall incorporate a minimum of one of the following materials conservation features. The product must be present in the entire application for which it is used.

<u>Insulation:</u>

(As a minimum use entirely in either the attic, the main walls or in the basement walls.)

#1   Fibre Glass Insulation
– a minimum of 50% of the raw material is recycled glass

#2   Cellulose Insulation
– product is manufactured from 100% recycled paper

#3   Mineral Fibre Insulation
– a minimum of 50% of the raw material is recycled

<u>Roofing:</u>

(Use over the entire roof.)

#4   Roofing System
– system contains compatible recycled materials (e.g., metal, paper, wood fibres, plastic or rubber)

<u>Sheathing/Drywall:</u>

(Material conservation product must replace equivalent "conventional" product throughout the house.)

#5   Fibreboard
– product is made from recycled newsprint and/or wood fibres

#6   Siding
– manufactured from factory and sawmill waste

#7   Drywall
– product contains recycled gypsum and/or newsprint

<u>Interior:</u>

(Must replace equivalent "conventional" product for entire floor.)

#8   Steel Studs
– a minimum of 60% of the raw material is recycled steel

#9   Studs and Trims
– manufactured from sawmill cut-offs and waste
– urea-formaldehyde free

<u>Exterior:</u>

(Must replace equivalent "conventional" backfill in its entirety.)

#10   Foundation and/or Under Slab Drainage
– mixture of rock and post consumer glass

Appendix V

BIBLIOGRAPHY and REFERENCES

The Canadian research and construction community has played a major role in advancing the principles and practices of energy-efficient and environmentally sustainable housing. The Canadian Home Builders' Association (CHBA), Canada Mortgage and Housing Corporation (CMHC) and the Canada Centre for Mineral and Energy Technology (CANMET), the technology development arm of Natural Resources Canada, have been responsible for a significant amount of background research and technology transfer initiatives. Their addresses, for information on publications and other resources, are included at the end of this Reference Section.

General Energy-Efficient Design

Building Science for a Cold Climate, N. Hutcheon and G. Handegord, John Wiley and Sons, (1983).

HOT2000: An Advanced Approach to the Design & Modelling of Energy Efficient Structures, CANMET (1993).

HOT2000 Program Comparison / Validation with US Blast 3.0 Computer Program, CANMET (1993).

Building Performance

The Flair Homes Energy Demo / CHBA Flair Mark XIV Project: Description of the Project Houses, Monitoring Program and Data Base, CANMET (1992).

Field Performance of Energy Efficient Residential Building Envelope Systems, CANMET (1992).

Measured Airtightness of 24 Detached Houses Over Period of up to Three Years, CANMET (1992).

Incremental Costs of Residential Energy Conservation Components and Systems, CANMET (1992).

Advanced Houses Program Technical Requirements, CANMET (1992).

Performance of the Brampton Advanced House, CANMET (1992).

Advanced Houses – Testing New Ideas, CANMET (1993).

Building Envelope Construction

Airtightness and Air Quality in Preserved Wood Foundations, CMHC (1992).

Building Successful Floor Systems: A Guide for Builders, CHBA/Ontario New Home Warranty Program (1994).

Research Project on the Noise Isolation Provided by Floor / Ceiling Assemblies in Wood Construction – Stage I, MJM Conseillers en Acoustique Inc., CMHC (1990).

Sound Performance of Wood Floor / Ceiling Assemblies, Stage II, William Bradley, CMHC (1990).

Method of Wooden Floor Construction for Minimizing Levels of Vibration, Y.H. Chui, CMHC (1991).

Testing of Air Barrier Construction Details, CMHC (1991).

Exterior Insulation and Finish Systems (EIFS) Problems, Causes and Solutions, CMHC (1993).

Air Leakage Characteristics of Various Rough-Opening Sealing Methods for Windows and Doors, CANMET (1992).

A Study of the Long-term Performance of Operating and Fixed Windows Subjected to Pressure Cycling, CANMET (1993).

A Preliminary Assessment of the Solar Shade Screen System for Reducing Residential Cooling Loads, CANMET (1992).

CSA–A440.2 "Standard for Energy Ratings for Windows and Sliding Doors", Canadian Standards Association.

Energy Saving Windows, CANMET (1993).

WIN–2000, Vision and FRAME Computer Programs, CANMET.

Mechanical Systems

CAN/CSA-F326-M91 "Residential Mechanical Ventilation Systems", Canadian Standards Association.

Earth Energy Systems: A Guide to the Technology, CMHC (1992).

Field Performance of Various Types of Residential Mechanical Ventilation Systems, CANMET (1992).

Study of Residential Ventilation Duct Energy Losses, CANMET (1992).

R-2000 Make-up Air Guidelines, CHBA (1993).

Home Automation Systems: An Overview for Canada, Canadian Automated Building Association (1992).

Environmentally Sustainable Housing

Builder's Guide to Environmental Products, CHBA (1994).

Clean Air Guide: How to Identify and Correct Indoor Air Quality Problems, CMHC, NHA 6695 (1994).

Healthy Housing: A Guide to a Sustainable Future, CMHC, NHA 6725 (1993).

Housing for the Environmentally Hypersensitive: Surveys and Examples of Clean Air Housing in Canada, CMHC (1993).

Making a Molehill Out of a Mountain: Reducing Residential Construction Waste, CMHC (1991).

Residential Water Conservation: A Review of Products, Processes and Practices, CMHC (1992).

Accessibility

Housing for Disabled Persons, CMHC, NHA 5467 (1993).

Housing Choices for Canadians with Disabilities, CMHC, NHA 6619 (1993).

Periodicals

Energy Design Update, Cutter Information Corp., 37 Broadway, Arlington, MA 02174

Environmental Building News, RR1, Box 161, Brattleboro, VT 05301, CT 06470

Fine Homebuilding, The Taunton Press, P.O. Box 355 — Newton, CT 06470

Home Builder Magazine, Work-4 Projects Ld., P.O. Box 400, Victoria Station, Westmount, Québec, H3Z 2V8

Northwest Builder, Iris Communications Inc., 258 East 10th Ave. Suite E, Eugene OR 97401

The Journal of Light Construction, P.O. Box 680, Mt. Morris, IL 61054

Home Energy: The Magazine of Residential Energy Conservation, 2124 Kitredge St. #95, Berkely, CA 94704

Solplan Review, The Drawing-Room Graphic Services, P.O. Box 86627, North Vancouver, B.C., V7L 4L2

Organizations and Agencies

Canadian Home Builders' Association
150 Laurier Ave. West, Suite 200
Ottawa, Ontario
K1P 5J4

Canada Mortgage and Housing Corporation (CMHC)
Canadian Housing Information Centre
700 Montreal Road
Ottawa, Ontario
K1A 0P7

Canada Centre for Mineral and Energy Techology
CANMET — Buildings Group
580 Booth Street, 9th Floor, Room 973
Ottawa, Ontario
K1A 0E4

Energy Efficient Building Association
Northcentral Technical College
1000 W. Campus Drive
Wausau, Wisconsin 54401 — 1899

Appendix **VI**

GLOSSARY

Adfreezing The process whereby wet soils freeze to below grade materials such as foundation walls or insulation, forcing movement of the material.

Air Barrier A material carefully installed within a building envelope assembly to minimize the uncontrolled passage of air into and out of a dwelling. Air barriers may be made of either membrane materials such as polyethylene or rigid materials such as drywall. In either case, they are carefully sealed to ensure continuity.

Air Change Per Hour (ach) A unit that denotes the number of times per hour that the entire volume of air in a house is exchanged with outside air. This is generally used in two ways: 1) under natural conditions and 2) under a pressure difference of 50 pascals. A well-built house should have no more than 1.5 ach at 50 pascals.

Air Leakage The uncontrolled flow of air through a component of the building envelope, or the building envelope itself, that takes place when a pressure difference is applied across the component. Infiltration refers to air leakage into the house, and exfiltration refers to the outward leakage of air.

Air Permeability A measurement of the degree to which a building component allows air to pass through it when it is subjected to a differential pressure.

Air Pressure The pressure exerted by the air. This may refer to static (atmospheric) pressure, or dynamic components of pressure arising from air flow, or both acting together.

Air Sealing The practice of sealing (from the interior) unintentional gaps in the building envelope in order to reduce uncontrolled air leakage. All joints, holes and cracks in the air barrier should be carefully sealed with compatible caulking materials or gaskets.

Airtightness The degree to which unintentional openings have been avoided in a building's structure.

Ambient Temperature The temperature of the air, (1) outside the building, or (2) within a room.

Automatic Flue Damper A damper added to the flue pipe downstream of a furnace or boiler and connected with automatic controls to the burner. Its function is to reduce heat loss up the chimney when the unit is not operating. Consequently, it provides the greatest savings during relatively mild weather when the furnace is operating only infrequently.

Backdrafting (flow reversal) The reverse flow of chimney gases into a building through the barometric damper, draft hood or burner unit. Backdrafting can be caused by chimney blockage, or it can occur when the pressure differential is too high for the chimney to draw.

Building Orientation The siting of a building on a lot. The term is generally used when discussing solar orientation, which is the siting of a building with respect to access to solar radiation.

Combustion Air The air required to provide adequate oxygen for fuel-burning appliances in a building.

Condensation The beads or drops of water (and frequently frost or ice in extremely cold weather) that accumulate on the inside of the exterior covering of a building (most often on windows). Condensation occurs when warm, moisture-laden air from the interior contacts a colder surface; the air cools and can no longer hold as much moisture.

Convection The transfer of heat from one point to another in a fluid (e.g., air, water), by the mixing of one portion of the fluid with another.

Controlled Ventilation Ventilation brought about by mechanical means through pressure differentials induced by the operation of a fan.

Dampproofing The process of coating the exterior of a foundation wall with bituminous emulsions or plastic cements. The purpose of dampproofing is to prevent or interrupt the capillary draw of moisture into the wall or floor system and to the interior of the foundation.

(De)humidistat An electronic sensing and control device used to regulate mechanical ventilation according to the relative humidity in the building. When the relative humidity surpasses a preset limit, the dehumidistat activates the ventilation system in order to exhaust house air.

Design Heat Losses The total predicted envelope heat losses over the heating season for a particular house design in a particular climate.

Dewpoint The temperature at which a given air/water vapour mixture is saturated with water vapour (i.e., 100% relative humidity). If air is in contact with a surface below this temperature, condensation will form on the surface.

Dilution Air The air required by some combustion heating systems in order to isolate the furnace from outside pressure fluctuations and to maintain a constant chimney draft.

Direct Gain A term referring to a type of solar heating system where the solar collection area is an integral part of the building's usable space; for example, windows.

Envelope The exterior surface of a building including all external additions, e.g., chimneys, bay windows, etc.

Equivalent Leakage Area (ELA) An estimate of the total area of all the unintentional openings in a building's envelope, generally expressed in square centimetres or square inches.

Exfiltration The uncontrolled leakage of air out of a building.

Expansive Soils Fine-grained soils such as clay, silt and fine sand which are frost-susceptible and which are therefore more subject to frost heaving.

Fan Depressurization A process by which a large fan is used to exhaust air from a building in order to create a pressure difference across the building envelope. An analysis of the flow rate through the fan at different pressure differences provides a measurement of airtightness.

Frost Heaving The movement of soils caused by the phenomenon known as ice lensing or ice segregation. Water is drawn from the unfrozen soil to the freezing zone, where it attaches to form layers of ice, forcing soil particles apart and causing the soil surface to heave.

Heating Degree Day The number of degrees of temperature difference on any one day between a given base temperature and the mean daytime outside temperature. The base is usually 18°C (64°F). The total number of degree days over the heating season indicates the relative severity of the winter for a specific location.

Heat Recovery The process of extracting heat (usually from a fluid such as air or water) that would otherwise be wasted. Heat recovery in housing usually refers to the extraction of heat from exhaust air.

Humidistat *See* Dehumidistat.

Ice Lensing *See* Frost Heaving.

Impermeable Not permitting water vapour or other fluid to pass through.

Induced Draft Flue System A term referring to a type of heating system equipped with a fan downstream of the furnace. The fan pulls gases from the furnace and propels them to the outside, thereby eliminating the requirement for dilution air.

Infiltration The uncontrolled leakage of air into a building.

Intrinsic Heat Heat from human bodies, electric light bulbs, cooking stoves, and other objects not intended specifically for space heating.

Latent Heat Heat added or removed during a change of state (for example, from water vapour to liquid water) while the temperature remains constant.

Mechanical Systems All the mechanical components of the building, i.e., plumbing, heating, ventilation, air conditioning and heat recovery.

Moisture Barrier A material, membrane or coating designed to keep moisture coming through the foundation wall from penetrating the interior foundation insulation or the wood members supporting the insulation and interior finishes.

Negative Pressure A pressure below atmospheric pressure. A negative pressure exists when the pressure inside the house envelope is less than the air pressure outside. Negative pressure will encourage infiltration.

Normalized Leakage Area (NLA) The NLA is calculated by dividing the Equivalent Leakage Area (ELA) from a fan test by the area of the exterior envelope of the house.

Orientation The direction (with respect to the points of a compass) in which the building axis lies or external walls face.

Pascal A unit measurement of pressure in the SI (metric) system. House airtightness tests are typically conducted with a pressure difference of 50 pascals between the inside and outside. 50 pascals is equal to 5.008 mm (0.2 in.) of water at 12.9°C (55°F).

Permeance Water vapour permeance is the rate at which water vapour diffuses through a sheet of any thickness of material (or assembly between parallel surfaces). It is the ratio of water vapour flow to the differences of the vapour pressures on the opposite surfaces. Permeance is measured in perms (ng/Pa.m²s, or grain/ft²hr (in.Hg)).

Positive Pressure A pressure above atmospheric pressure. A positive pressure exists when the pressure inside the house envelope is greater than the air pressure outside. A positive pressure difference will encourage exfiltration.

Pressure Difference The difference in pressure of the volume of air enclosed by the house envelope and the air surrounding the envelope.

Radiant Heat Transfer The transfer of heat energy from a location of higher temperature to a location of lower temperature by means of electromagnetic radiation.

Relative Humidity The amount of moisture in the air compared to the maximum amount of moisture that the air could retain at the same temperature. This ratio is expressed as a percentage.

Resistance Value (RSI) Thermal resistance value. This is a measurement of the ability of a material to resist heat transfer.

Sealants Flexible materials used on the inside of a building to seal gaps in the building envelope in order to prevent uncontrolled air infiltration and exfiltration.

Solar Heat Gain A term used in the passive solar heating field to describe the amount of heat gained through windows during the heating season. Net solar gain refers to the solar heat gain less the heat losses through the windows.

Stack Effect Pressure differential across a building caused by differences in the density of the air due to a difference of temperature between indoors and outdoors.

Thermal Break A material of low conductivity used in a building assembly to reduce the flow of heat by conduction from one side of the assembly to the other. The term is often used to refer to materials used for this purpose in the frame of metal windows.

Vapour Diffusion The movement of water vapour between two areas caused by a difference in vapour pressure, independent of air movement. The rate of diffusion is determined by (1) the difference in vapour pressure; (2) the distance the vapour must travel; and (3) the permeability of the material to water vapour (hence the selection of materials of low permeability for use as vapour diffusion retarders in buildings).

Vapour Diffusion Retarder Any material of low water vapour permeability used to restrict the movement of water vapour due to vapour diffusion. There are two classes of VDRs. A *Type I* VDR has a permeance of 14.375 metric perms (0.25 perms in imperial units) or less. A *Type II* VDR has a permeance of 43.125 (0.75) perms or less before aging and 57.5 (1.0) perms or less after aging. Any material with a perm rating higher than 57.5 (1.0) perms is *not* a VDR.

Water Vapour Pressure The pressure exerted by water vapour in the air. Water vapour moves from an area of high pressure to an area of low pressure.

INDEX

NOTES